DEVELOPMENTS IN MARITIME TRANSPORT AND LOGISTICS IN TURKEY

T0222090

Developments in Maritime Transport and Logistics in Turkey

Edited by
MAHMUT CELAL BARLA
Department of Maritime Transport Business, Istanbul University
OSMAN KAMIL SAG
Maritime Faculty, Istanbul Technical University
MICHAEL ROE AND RICHARD GRAY
Institute of Marine Studies, University of Plymouth

Routledge
Taylor & Francis Group

LONDON AND NEW YORK

First published 2001 by Ashgate Publishing

Published 2017 by Routledge
2 Park Square, Milton Park, Abingdon, Oxfordshire OX14 4RN
711 Third Avenue, New York, NY 10017, USA

First issued in paperback 2017

Routledge is an imprint of the Taylor & Francis Group, an informa business

British Library Cataloguing in Publication Data
Developments in maritime transport and logistics in Turkey.
    - (Plymouth studies in contemporary shipping)
    1.Shipping - Turkey 2.Logistics, Naval
    I.Barla, Mahmut Celal
    387.5'09561

Library of Congress Control Number: 2001089776

ISBN 13: 978-1-138-26378-9 (pbk)
ISBN 13: 978-0-7546-1392-3 (hbk)

Contents

Acknowledgements

Our greatest thanks go to a variety of people who have been indispensable in the development of this text. To our friends and close colleagues in Turkey - Güldem Cerit, Funda Yercan and Okan Tuna at Dokuz Eylul University; Ozgul Oney at Shell Bunkers, Istanbul; and the substantial numbers of students who have passed through the University of Plymouth from Turkey and contributed so much to the courses and research life of this city. To Marie Bendell at the Institute of Marine Studies, without whose support there would be no research output, just chaos. And to all our families for their continued trust and faith.

1 Introduction

MICHAEL ROE
INSTITUTE OF MARINE STUDIES
UNIVERSITY OF PLYMOUTH

This book is a result of a number of years of collaboration that has taken place between the University of Plymouth in the UK and Dokuz Eylul University in Izmir, Turkey. Through the close relationships that those in Izmir have developed through other universities in Turkey, a number of papers have been brought together which are of some significance to the development of the maritime economy in Turkey and which also reflect the standing of Dokuz Eylul University as the major research and teaching institution in the region for maritime business. In addition, the work of the other editors of this book needs to be recognised - Professor Mahmut Celal Barla of Istanbul University and Professor Osman Kamil Sag of Istanbul Technical University and Dr Richard Gray at the University of Plymouth - without whose close co-operation the text would never have emerged.

The Institute of Marine Studies at the University of Plymouth remains the leading research and teaching university for maritime business in Western Europe and retains its long-standing interests in Turkey as an educational and maritime partner. There has been a history of student and staff exchanges, collaborative research projects and combined publications through the 1990s fostered by the encouragement enthusiasm and academic leadership of Professor A. Güldem Cerit at Dokuz Eylul, without whom little would have been achieved. Additional thanks must also be passed to Dr H. Funda Yercan - a long-standing colleague and friend - and Dr Okan Tuna, both also from Dokuz Eylul, without whom the task would have been made that much more difficult.

The book itself ranges widely over Turkish maritime affairs reflecting the breadth of activity within that country in the shipping, ports and ancillary sectors.

The opening paper by Kaptanoglou and Roe was completed whilst the former was researching at the University of Plymouth and focuses upon the importance and development of the shipping industry in Turkey during the early and mid 1990s, a period of significance for the industry. It provides an interesting and detailed background for the papers that follow.

Tuna's detailed examination of service quality in container shipping, carried out in Turkey is a development of his doctoral research that was completed in Turkey and gives a unique interpretation of marketing issues in this sector of business which has been largely ignored from this perspective.

Meanwhile, the paper by Alkan, Cerit and Barla provides our first look at the ports of Turkey, and that of Mersin in particular, in terns of its development as a centre and stimulus for regional activity, examining its feeder activity in the container market place.

We next turn to legal matters where the paper by Baser and Baser takes a detailed look from a Turkish perspective, of the difficulties and specific needs of a maritime policy for straits with examples from the Mediterranean region. In so doing it represents a unique analysis, providing for the first time an attempt at bringing together a discussion that is particularly pertinent in the light of the problems presented by the Bosphorous Strait in Turkey.

Another Turkish perspective is taken by Kisi but this time in considering crisis management approaches in Turkish ports. The specific characteristics of port crisis management is considered here with examples form Turkey to illustrate the difficulties and requirements. This is followed by the paper from Cerit and Sag which turns to the important shipbuilding markets in Turkey and the impact of technological developments.

The importance of the Caspian Sea oil industry and the growth of exports from this region is examined by Yuceer and Cerit in their paper. Turkish shipowners are hopeful of playing a major role in this market in the future as the oil starts to flow from the new fields being discovered. Plans to develop pipeline connections to Turkish ports from the Caspian remain unresolved but offer the prospect of enormous potential markets for oil shipping in the future.

Shipbuilding is returned to in the paper by Bayraktar, Heijveld and Roe which provides a detailed examination of the development of the industry during the 1990s when its growth and technical expansion was notable. This paper is the result of a period of research carried out at the Institute of Marine Studies during the 1990s as part of a collaborative research programme.

Acar provides a Turkish perspective on reliability, availability, maintenance and safety in the maritime sector, taking a Turkish perspective and is valuable in considering these serious issues together in a rare example of research in this sector. Alkan, Bak and Cerit review the role and importance of ergonomics in shipping using examples from the Turkish industry, whilst Devici, Cerit and Sigura analyse the role of liner agents in port service quality from the point of view of the container sector in a paper that provides a unique analysis of this industry.

Finally, Yercan covers the issues which are emerging from the continued privatisation of major ports in Turkey and the difficulties and opportunities that are being faced as this process continues.

In this way, this text provides coverage of the entire maritime sector in Turkey, from shipping to ports, shipbuilding to agencies, ergonomics to law in a way that has not been achieved before. It is hoped that the reader will enjoy discussion of the major issues in maritime affairs in Turkey as a result, as that country continues to integrate with the rest of Europe and its shipping, shipbuilding and ports sectors continue to grow and prosper.

Michael Roe
Plymouth, January 19th 2001.

2 The Role of Shipping in the Turkish Economy in the 1990s

MEHMET KAPTANOGLOU AND MICHAEL ROE
INSTITUTE OF MARINE STUDIES
UNIVERSITY OF PLYMOUTH

Introduction

Towards the year 2000, the new order of the changing world has a tendency to move to increasing globalisation with many international activities developing closer relationships and links. Turkey is involved in this globalisation with the most recent example of the Customs Union with the European Union, which commenced on January 1, 1996.

Turkey has considerable potential for developing trade. One of the sources of this potential is the entrepreneurial nature of the private sector. The total amount of exports and volume of foreign trade has increased largely because of the liberalisation of the economy of the country during the 1980s. Considerable development in the Turkish shipping industry has been seen as a result of the growth in the economy.

The relationship between the sea and the Turkish people has been continuous since the 15th century. Although there has been a large Turkish fleet throughout history, the majority has been related to military and not mercantile objectives. In more recent years, the capacity of the Turkish merchant fleet has been increasing since the end of the 1980s. The total capacity of the fleet has doubled during this period and therefore, the role of Turkish shipping in the national economy has increased significantly in importance.

Although most industries in the country were affected by the economic crisis of 1994, the Turkish shipping industry largely escaped. The main reasons were the international activities of the shipping industry and its ability to access sources of foreign exchange.

This paper goes on to examine the relationship between the Turkish

4

shipping industry and the Turkish economy during the mid-1990s.

Turkey and its Economy

Background to Turkey

Turkey is a country with some very individual characteristics compared with most countries in the world. For example, Turkey has lands both in Europe and in Asia, Turkey is a Moslem country but has a liberal economy, and Turkey has an historical background based on the Ottoman Empire but remains to a large extent a developing country.

Turkey is situated on both the European and Asian continents with land totalling approximately 780,000 sq. km. The mainland is located on the south-west of the Asian continent, known as the peninsula of Anatolia. The European part of the country is rather small compared to the Asian part and it is situated in the far south-east of the continent.

There are four seas surrounding Turkey. These are the Black Sea to the north, the Marmara Sea to the north-west, the Aegean Sea to the west and the Mediterranean Sea to the south. The total length of coastline of Turkey is approximately 8,000 kilometres. There are two straits to the north-west of Turkey where the two continents face each other - the Bosphorus and the Dardanelles. The Black Sea to the north connects the Aegean Sea to the west through these straits, whilst the Marmara Sea is situated between them.

The neighbouring countries of Turkey are Iraq and Syria on the south-eastern border, Iran on the eastern border, Georgia and Armenia on the north-eastern borders and Greece and Bulgaria on the north-western borders.

The population of Turkey is approximately 60 million (1998). The number of inhabitants per square kilometre is 80. The annual growth rate for the population has been approximately 20% up until 1995 (Institute of Statistics, 1995).

The political nature of Turkey has tended to be to the right of the spectrum. The political characteristics of Turkey have generally been liberal with the main target of a liberalised economy dominating at present. There has been a coalition in power from the beginning of 1996, between parties from the right side.

Turkey is both a developing and an industrialising country. Approximately 65% of the population lives in urban areas. The annual growth rate of these urban areas was until the mid 1980s, approximately 40% whilst the annual growth rate of rural areas was approximately -6% (Middle East Research Institute, 1985). These percentages clearly indicate that there has been a rapid movement of people from rural to urban areas largely caused by the sizeable development in industry and education.

Economic Structure

The economy was closed and dependent upon domestic resources until the beginning of the 1980s. After the Prime Ministry of Ozal in 1983, the Turkish economy was largely liberalised (Middle East Research Institute, 1985). Since then, further steps have been taken towards extending the economic development of Turkey including in particular, its further industrialisation.

The characteristics of the current Turkish economy have been derived from the liberalisation policies developed during the Prime Ministry of Ozal through the 1980s. Liberalisation within the economy has played an important role for Turkey and the shipping industry has been one economic activity that has been particularly affected.

There was considerable instability and uncertainty within the Turkish economy in the period leading up to the 1980s severely affecting many economic sectors. However, the situation began to recover after the military action of 1980. Democracy again took its place in the political arena in 1983. Turgut Ozal, with a background from the USA, was elected as the Prime Minister from the Motherland Party. This was a turning point for the economy (and society) which has had substantial implications for the shipping industry amongst others.

The major changes introduced by the Ozal government during this period included the following:

- the Turkish Lira has become a convertible currency;
- limitations on the possession of foreign currency have been removed;
- stock exchange markets have been extensively developed and become considerably more active;
- exports and imports have become easier;

- bureaucratic problems have been reduced;
- a variety of subsidies have been given to private entrepreneurs to increase exports.

As a result, the Turkish economy showed sporadic signs of recovery and exports rose to 7.7 billion US$ in 1985 from 2.5 billion US$ in 1981. The rise of GNP during this period was approximately 5.2%. Before the 1980s, Turkey was a country whose economy was dependent upon agriculture. However, during the era of Ozal, and after the 1980s, the percentage production based upon the industrial sector has overtaken that based upon agriculture. This increase has also continued throughout the 1990s. Figure 1 illustrates the volume of export foreign trade for Turkey for various sectors, in 1993 and 1994.

Figure 1: Export Foreign Trade Volume by Sector (mn US$) (1993-94)

	1993	1994	% increase
Agricultural products	2.381	2.470	+3.7
Minerals	2.380	2.720	+14.2
Industrial products	12.726	15.364	+20.7
Total	**15.345**	**18.106**	**+11.4**

Source: State Institute of Statistics, 1995.

The Present Condition of the Turkish Economy

There has been a severe economic crisis in Turkey since 1994 although this is in contrast with many other countries in the world. The annual inflation rate in Turkey reached 150% in 1995 and the budget deficit was also notably increasing. The volume of economic activity decreased by 6% during the same year.

The pattern of foreign trade in Turkey in the mid-1990s, reflected a substantial growth in the difference between imports and exports. Figure 2 illustrates some foreign trade indicators for Turkey between 1992 and 1994.

Figure 3 illustrates Turkish foreign trade volumes according to the value of imports and exports in million US$ for the period between 1985 and 1995. Although, the volume of imports and exports appears to have doubled between 1985 and 1995, the volume actually decreased between 1993 and 1994 before rising again. Figure 4 illustrates the amount of imports and exports in tons.

Figure 2: Foreign Trade Indicators for Turkey 1992-4 (million US$)

	1992	1993	1994
Imports	22.871	29.428	23.270
Exports	14.714	15.345	18.106
Foreign trade volume	37.588	44.773	41.376
Trade deficit	8.157	14.083	5.164
Export/Import	64%	52%	78%
Export/GNP	55%	92%	29%
Import-export/GNP	24%	25%	31%

Source· State Institute of Statistics, 1995.

Figure 3: Turkish Foreign Trade Development by Value (mn US$)
          (1985-95)

	Imports	Exports
1985	11.343	9.958
1986	11.104	7.456
1987	14.157	10.190
1988	14.335	11.662
1989	15.792	11.624
1990	22.302	12.959
1991	21.047	13.593
1992	22.871	14.715
1993	29.428	15.345
1994	23.270	18.106
1995	25.708	21.636

Source. State Institute of Statistics, 1995.

By the end of 1994, the economy had started to recover and the volume of foreign trade again began to increase with a number of European countries. Figure 5 illustrates the foreign trade volumes between Turkey and various countries.

The foreign trade volumes estimated for 1995 were more than achieved. The estimated exports were 19.5 million US$, but the actual amount reached 21.5 million US$. Similarly, estimated imports were 27 million US$, but the actual amount achieved was 32 million US$. However, the deficit between imports and exports also increased in 1995.

Figure 4: Turkish Foreign Trade (thousand tons)

	Imports	Exports
1985	33.329	13.789
1986	35.592	13.221
1987	45.552	14.261
1988	46.304	21.891
1989	47.168	17.421
1990	50.887	18.649
1991	47.607	23.741
1992	50.573	26.051
1993	65.497	21.604
1994	56.892	28.578

Source: State Institute of Statistics, 1995.

Figure 5: Foreign Trade Volumes Between Turkey and Various European Countries in 1995 (million US$)

	Export	% Share	Import	% Share
Germany	5.036	44.1	5.548	30.2
USA	1.514	13.3	3.724	20.3
Italy	1.457	12.8	3.193	17.4
UK	1.136	10.0	1.830	10.0
France	1.033	9.1	1.996	20,9
Russia	1.238	10.8	2.082	11.3

Source. State Institute of Statistics, 1995.

The main economic problems for Turkey had stemmed from the economic structure and system that had been adopted. The main problems identified within the Turkish economy can be listed as follows:

- high inflation;
- imbalance of revenue and expenditure (with too much emphasis upon the latter);
- increasing discrepancies between imports and exports;
- investments that were not financially credible;
- the need to privatise state-owned institutions.

Turkish Shipping

Shipping in Turkey has played the role of a locomotive industry for the country and continues to play a significant role within the economy. The changes occurring within Turkish shipping since the beginning of the 1980s continued throughout the 1990s.

These changes are gaining momentum in the industry (Seatrade, 1995) and the volume of trade by sea has been increasing from year to year. Overall growth within the shipping industry is increasing and the Turkish merchant fleet has continued to expand.

Recent developments in the Turkish shipping industry are outlined in the following two sections for the periods during the 1980s and the 1990s.

The 1980s

Until the beginning of the 1970s, the shipping industry in Turkey was of little significance to the economy as a whole. Vessels were commonly only of low tonnage (up to 700 dwt).

Many steps were taken within the shipping industry, during the 1970s, to encourage its development. The Turkish Chamber of Shipping recently noted that Turkish shipping began to promote its activities, especially in the Mediterranean Sea, and the industry became highly profitable through its involvement in international trade during the mid 1970s. However, the Marine Bank of Turkey was declared bankrupt during this period and the Government no longer granted loans to the shipping industry. Some loans for relatively small amounts were distributed by various private banks and international finance institutions to a limited number of shipping companies for a narrow range of very specific projects.

The situation for the shipping industry in Turkey changed during the 1980s along with many sectors of the economy with the changing context and plans of the Prime Ministry of Ozal.

Both loans and subsidies for the maritime industry became available again and the Turkish Merchant Fleet reached approximately 2 million dwt by 1982. Self-reliance and the credibility of Turkey within the international markets of the world increased and the doors of the country were opened wide to international markets. As a result of this, international activities increased in many industries in Turkey and shipping was a major part of this. 86% of the total amount of exports and imports and 52% of the total value of exports and imports of Turkey were carried by sea transport during the beginning of the Ozal era. It is widely considered that shipping played an important role in the development of foreign trade of the country with the help of the newly liberalised economy.

The capacity of the Turkish Merchant Fleet reached 5.8 million dwt at the end of 1985. However, the total capacity of the fleet declined to 4.9 million dwt at the end of 1988 for various reasons. One was that the average age of the fleet was 19 years, which was high if compared to the world (Turkish Maritime Organisation, 1989). Thus many vessels had to be scrapped. Another reason was the world crisis in the shipping industry between 1986-1988. Turkish shipping was affected by that crisis. The total capacity of the fleet had recovered a little to reach 5.17 million dwt at the end of 1989.

The 1990s

The Turkish merchant fleet was progressively ageing during the early 1990s and vessels considered to be relatively new, were now nearly 10 years old. Meanwhile, the average age of the Turkish merchant fleet by 1991 was 21 years, and as such was one of the oldest fleets in the world (Ministry of State, 1992). There was thus an urgent need for the Turkish fleet to be renewed before its efficiency began to be compromised.

By 1992 the total capacity of the Turkish merchant fleet was some 6.5 million dwt. Of this, the total capacity of the fleet owned by the state was around 1.3 million dwt representing around 20% of the total Turkish merchant fleet. On the other hand, the total capacity of the fleet belonging to the private sector was around 5.2 million dwt representing the remaining 80%. This placed it at 30th position in the world in 1992, representing 1% of the world merchant fleet (Tez, 1993).

At this time the Turkish fleet was expanding rapidly and both in size and quality. The total capacity of the fleet was 7.93 million dwt in September 1993 and was 8.25 million dwt by the end of 1993 (Seatrade, 1993).

Prime Minister Demirel from the True Path Party, stepped into the Presidential position after the sudden death of President Ozal. Professor Ciller, who was the Minister of State with responsibility for the economy, was elected as the Head of the True Path Party and became the first woman Prime Minister of Turkey, in 1993.

During the Prime Ministry of Ciller, Tez was the Minister of State with responsibility for maritime affairs. Although Tez had no maritime experience or background, he had strong links with the sector and the Chamber of Shipping and he had conducted many discussions concerning maritime policy issues during his political progress. During the Ciller Prime Ministry, an undertaking was given that an Undersecretariat for Maritime Affairs would be established, directly under the Prime Ministry. Additionally, the Ministry of Maritime Affairs would be established in the near future (Seatrade, 1993). In fact this was not to occur until the year 2000 (Lloyd's List, 2000).

The rapid growth in the merchant fleet brought many problems including for example, those associated with the training of both seafarers and their managers. As a result of the inadequacies in training personnel employed in the maritime sector, many deficiencies in the industry occurred, and due to the increasing time spent in correcting errors, running costs undoubtedly increased (Lloyd's Ship Manager, 1993).

There were also problems with the Turkish economy in 1993. The country faced a very high inflation rate of approximately 80% annually. Increasing investment and subsidies were needed by the shipping industry to overcome the problems exacerbated by these economic difficulties.

The first half of 1994 was also traumatic for the Turkish economy. A very high rate of inflation and increasing interest rates were experienced. Annual inflation reached an average of 150% and at times it even exceeded 1,000%. As a consequence, a collapse in the international value of the Turkish lira occurred. This economic crisis affected the large majority of sectors in the country and even though shipping was perhaps less affected because of its international nature, the commodity base upon which it depended was markedly depressed. As a result of this economic instability, an emergency economic austerity package was adopted by the government at the beginning of 1994, in an attempt to restore stability to the economy.

The total Turkish merchant fleet increased to 9.5 million dwt in 1994 with some 1,025 vessels. In terms of dwt, almost half of the total fleet comprised bulk carriers, although in vessel number terms the biggest category was still general cargo vessels. Oil tankers increased in relative proportion. However, there were still at this stage no Turkish owned fully cellular, container ships, although some multi-purpose vessels were then being used to carry containers.

Secondhand vessel purchases reflected the age profile of the fleet and reinforced the fact that over half of the fleet (around 5.8 million dwt) had ages between 15 and 24 years.

Turkish Shipping in the mid-1990s

The shipping industry has been for many years, one of the leading industries in Turkey playing a significant role in the economy of the country. There is an approximate movement of 4.5 billion tons of goods in the world each year by sea and the world merchant fleet consists of around 700 million dwt. Sea transportation is approximately seven times cheaper than land transportation on average and is approximately 3.5 times cheaper than rail transportation - where this is practical and realistic. As a result, some 90% of the total transport in the world is conducted by maritime transport. In contrast to this, only 85% of total Turkish transport is the responsibility of maritime transport.

The total capacity of the Turkish merchant fleet reached 10.56 million dwt at the end of 1997, up from 4.11 million dwt recorded at the end of 1982 representing a notable fleet expansion in tonnage during this period. Figure 6 illustrates the development of the Turkish merchant fleet in dwt between 1982-1997.

The total number of vessels in the fleet at the end of 1997 was 1,197 with an average age of over 17 years. 91.5% of the fleet belonged to the private sector with the remainder belonging to the state. 66% of the vessels were bulk carriers, 9% oil tankers, 14% dry cargo vessels, 6% OBO, 2% ro-ro and ferries and 3% others. The Turkish merchant fleet by the end of 1997 was 17th in size in the world representing some 1.4% of the total tonnage (Yercan and Roe, 1999). Figures 7a and 7b indicate the vessel types within the Turkish merchant fleet at the end of 1995 - the latest year for which accurate data is available.

Within the Turkish fleet, the age distribution at the end of 1997 was characterised by 80 vessels between 0 and 4 years, 115 vessels between 5 and 9 years, 190 vessels between 10 and 14 years, 191 vessels between 15 and 19 years, 235 vessels between 20 and 24 years, 151 vessels between 25 and 29 years, and 235 vessels of 30 years or more. The large majority of vessels were thus over 15 years of age indicating a relatively old fleet (Chamber of Shipping, 1998). Figure 8 indicates the age groupings of the Turkish merchant fleet at the end of 1997.

The total Turkish carriage of goods by sea between 1985-1997 is illustrated in Figure 9. Carriage is distributed between Turkish and foreign flag vessels operated by Turkish operators.

Both passenger and freight vehicle ferries operate on domestic and international lines. The main operator is Turkish Maritime Lines which until 2000, was a part of the state-owned Turkish Maritime Organisation. There are also a number of other operators from the private sector.

Figure 6: Development of the Turkish Merchant Fleet 1982-1997

Year	Turkish Fleet (million dwt)
1982	4.11
1983	4.86
1984	6.05
1985	5.81
1986	5.23
1987	5.24
1988	4.91
1989	5.17
1990	5.64
1991	5.96
1992	6.50
1993	8.25
1994	9.50
1995	10.31
1996	10.89
1997	10.56

Source: Chamber of Shipping, 1998.

Figure 7a: Turkish Merchant Fleet (1995) by Vessel Number

	Imports	Built	Total
Dry cargo	107	368	475
Ro-ro	22	0	22
Bulk	151	6	157
Container	1	2	3
OBO	9	0	9
Tanker	67	88	153
Other	94	229	323
Total	451	691	1,142

Source: Sector Report on Turkish Shipbuilding Industry, 1995.

Figure 7b: Turkish Merchant Fleet (1995) by dwt

	Import	Built	Total
Dry cargo	582,437	873,750	1,456,187
Ro-ro	128,442	0	128,442
Bulk	5,667,527	92,337	5,759,864
Container	4,140	9,655	13,795
OBO	1,040,934	0	1,040,934
Tanker	1,668,917	153,773	1,822,690
Others	33,348	55,101	88,449
Total	9,125,745	1,184,616	10,310,361

Source: Sector Report on Turkish Shipbuilding Industry, 1995.

Figure 8: Age Grouping of the Turkish Merchant Fleet (end 1997)

Age	Number	Dwt (million)
0-4	80	0.35
5-9	115	0.39
10-14	190	2.03
15-19	191	2.08
20-24	235	3.76
25-29	151	1.60
30 +	235	0.35
Total	1,197	10.56

Source: Chamber of Shipping, 1998.

Turkish cabotage shipping interests consist of the carriage of oil, oil products, iron ore, coal, wheat, general cargo and other cargo. Cabotage reached its recent maximum amount in 1989, but has declined since then and this trend continues. The total cabotage carryings on Turkish domestic lines were 15,357,566 tons for loading and 18,646,496 tons for unloading in 1994. Figure 10, illustrates cabotage carryings on the domestic lines during the years 1987-1994.

Turkish Shipping on International Lines

Turkish international maritime trade consists of loading goods for export, unloading goods for import, servicing transit goods, and ro-ro and ferry operations.

Figure 9: Turkey - Carriage of Goods by Sea 1985-1997 (million tons)

Year	Turkish vessels	Foreign vessels	Total
1985	18.3	22.9	41.2
1986	17.7	24.7	42.4
1987	21.0	27.5	48.5
1988	19.7	32.8	52.5
1989	20.6	34.6	55.2
1990	22.3	36.8	59.1
1991	22.7	47.5	70.2
1992	29.5	42.9	72.4
1993	33.5	49.5	83.5
1994	37.0	37.8	74.8
1995	35.2	49.0	84.2
1996	36.1	55.6	91.7
1997	32.8	79.5	112.3

Source: Chamber of Shipping, 1998.

Trade activities in the Black Sea region have been increasing substantially following the collapse of the Soviet Union in 1991 although in most recent years since 1998, the problems of the Russian economy have interrupted progress. A trade agreement was signed between the countries of the region in 1992 - the Economic Co-operation Agreement for the Black Sea Region. The trade zone between these countries is termed the Black Sea Economic Co-operation Zone. Trade between Turkey and Azerbaijan, Georgia, Russia, Moldavia and Romania had risen by 40% in 1992 reaching 1.7 billion US$ and still continues to rise.

The total tonnage of trade between Turkey and the other countries with coasts on the Black Sea, including Ukraine, Bulgaria, Romania, Armenia, Azerbaijan, Georgia and Moldavia, has reached approximately 14 million tons of goods annually since 1995.

There are also substantial passenger and vehicle carryings on Turkish international lines carried on either ferries or ro-ro vessels. Lines operate to countries such as Romania, Ukraine, Russia and Azerbaijan in the Black Sea, to Italy, Cyprus, Libya and Israel in the Mediterranean Sea and various small vessels to the Greek islands in the Aegean Sea.

The main sectors within the Turkish merchant fleet consist of bulk carriers, oil tankers, dry cargo carriers, OBOs, ro-ro vessels and ferries, whilst other types include chemical carriers, container carriers, train ferries, livestock carriers, fishing boats, tugboats and research vessels.

Figure 10: Cabotage in Turkey, 1987-1994 (million tons)

Year	Oil/oil products	Iron ore	Coal	Wheat	General cargo	Total
1987						
Loaded	16.5	1.2	1.5	0.2	1.9	**21.3**
Unloaded	15.5	1.2	1.5	0.3	6.8	**25.3**
1988						
Loaded	18.4	1.3	1.8	0.2	2.0	**23.7**
Unloaded	17.3	1.2	2.1	0.2	3.1	**23.9**
1989						
Loaded	19.3	1.8	1.6	0.6	2.7	**26.0**
Unloaded	18.7	1.6	1.5	0.6	8.4	**30.8**
1990						
Loaded	14.6	1.4	1.4	0.8	2.3	**20.5**
Unloaded	14.8	1.3	1.3	1.0	8.3	**26.7**
1991						
Loaded	9.6	0.9	1.1	0.2	2.2	**14.0**
Unloaded	9.3	0.9	1.1	0.2	6.8	**18.3**
1992						
Loaded	9.7	0.7	1.2	0.4	3.1	**15.1**
Unloaded	9.6	0.8	1.3	0.3	7.2	**19.2**
1993						
Loaded	10.6	0.7	1.1	0.1	2.9	**15.4**
Unloaded	10.1	0.7	1.1	0.2	8.7	**20.8**
1994						
Loaded	10.4	0.7	0.9	0.1	3.3	**15.4**
Unloaded	10.4	0.6	0.9	0.1	6.6	**18.6**
Total						
Loaded	**109.1**	**8.7**	**10.6**	**2.6**	**20.4**	**151.4**
Unloaded	**105.7**	**8.3**	**10.8**	**2.9**	**55.9**	**166.6**

Source: Chamber of Shipping, 1995.

The total number of vessels within most sectors of the Turkish merchant fleet has been increasing since 1985. The total number increased 31% between 1985 and 1994. The increase has been particularly in the sectors of water tankers, OBO vessels, chemical carriers, ro-ro vessels and ferries, bulk carriers and fishing vessels.

There has been a decline in the total number of vessels belonging to the state in the sectors of bulk carriers, dry cargo carriers and tankers. The increased development of foreign trade transport of Turkey by sea is mainly distributed amongst the sectors of dry bulk, liquid bulk and general cargo. Growth cargoes are grain, metal ore, coal, oil and general cargo. The distribution of these cargoes is illustrated in Figure 11 for the years 1985, 1990 and 1994.

Problems with Turkish Shipping

The Turkish merchant fleet, which had a total capacity of approximately 5.17 million dwt at the end of the 1980s, reached a capacity of 10.31 million dwt at the beginning of 1996, and 10.56 dwt by the beginning of 1998. If the total capacity of bareboat charters is included in this, then the total capacity of the Turkish merchant fleet becomes approximately 11 million dwt at the beginning of 1996 and 12 million dwt by 1998.

Figure 11: Distribution of the Sectors of Turkish Foreign Trade by Sea Transport in 1985, 1990 and 1994 (million tons)

	1985	1990	1994
Grain	1.0	3.0	2.0
Metal ore	5.0	5.0	5.0
Coal	3.0	5.0	6.0
Oil	15.0	20.0	30.0
General cargo	20.0	25.0	35.0

Source. Chamber of Shipping, 1995.

Although the Turkish merchant fleet continues to grow, the industry still has many problems that need to be resolved in the relatively short term if it is to prosper.

A conference held in Kocaeli, in Turkey, in April 1996, concentrated upon the main problems for Turkish shipping. Although it was noted that the Turkish merchant fleet had grown rapidly after the liberalisation phase within the economy in the 1980s, the main problems still facing the fleet were classified into the following eight groups:

1) the shipping industry needed to increase its efficiency and more importance needed to be given to this industry within Turkey;

2) the mismatch in total carrying capacity of the Turkish merchant fleet compared with cargoes carried;

3) privatisation of Turkish Cargo Lines, various shipyards and

ports operated by the state was an urgent priority;
4) traffic through the two straits of the Marmara Sea, by which the Black Sea and the Aegean Sea are connected, presented many difficult and pressing issues;
5) incompatibility between Turkish domestic maritime issues and international maritime issues;
6) improvements needed in maritime education and training;
7) the establishment of a Ministry of Maritime Affairs;
8) severe financial problems within the industry.

More details of these points are explained below:

1) Efficiency within the Turkish shipping industry This problem is related to the ageing of the fleet, its average age compared with the world merchant fleet, standardisation within the Turkish merchant fleet and attempting to bring domestic standards up to those of the world merchant fleet. The total capacities of Turkish shipyards are considered to be too great because of inefficiency and low productivity. There are also technological problems at the ports. Another problem is the shortage of finance to solve these problems.

2) The total carrying capacity of the Turkish merchant fleet. The cargo carried by the Turkish fleet has always been below the actual capacity of the fleet. The total percentage of the utilisation of capacity of the fleet for foreign trade was 32.3% in 1991. This percentage increased to 49.5% in 1994 but it is still considered is not enough. Additionally, this percentage again decreased in 1995 to around 37%.

3) Privatisation. There have been many discussions related to privatisation of Turkish shipping over recent years and many promises have been given to the industry by politicians. However, little has actually been achieved. It is widely agreed that the eight major, state controlled ports and the various state operated shipyards need to be privatised (Lloyd's List, 1993). Additionally, Turkish Cargo Lines, operating bulk and general cargo vessels, is on the list awaiting privatisation moves. The state operated shipyards and ports lack both productivity and efficiency as a result of excessive bureaucratic procedures and inadequate management and decision-making.

4) Traffic through the two straits of the Marmara Sea. There have been many collisions and accidents while passing through the two straits of the Marmara Sea that connect the Black Sea to the north and the Aegean Sea

to the west of Turkey. Therefore, pressures have built up for an international agreement for a traffic control system in the increasingly busy Turkish straits, especially the Bosphorus strait in Istanbul. Some recent progress has been made here.

5) The incompatibility of many Turkish domestic maritime issues with international maritime issues and standards - which clearly causes some serious inconsistencies.

6) Maritime education and training. The Turkish fleet continues to grow and therefore, a substantial increase in maritime training and education is needed. There are a number of maritime schools and faculties in Turkey, but they are not sufficient for either seamanship training needs or for the management sector.

7) The Ministry of Maritime Affairs. There have been many promises given by politicians in Turkey for the establishment of a Maritime Ministry. Currently there is an Undersecretariat for Maritime Affairs, which is directly under the responsibility of the Prime Ministry. Some progress was announced in 2000 (Lloyd's List, 2000).

8) The financial situation presents further problems in dealing with all these issues. There has long been financial problems in the shipping industry. These problems exist because of the continuing difficult economic situation of Turkey, which has continuously interrupted subsidies by the Government and loans from the banks. Since there are no specialised banks related to the shipping industry in Turkey, financial support has to be taken from international banks and foreign sources.

The Role of Turkish Shipping in the Turkish Economy

The Turkish shipping industry has been one of the foundation stones of the economy of Turkey. The industry has developed substantially since the beginning of the 1980s and has contributed to the national economy through imports and exports of foreign trade.

Regarding the indicators of the State Institute of Statistics for 1994 (the most recent available), Gross National Product calculations were made for the main kinds of economic activity. The main GNP sectors for Turkey are agriculture, industry, construction and services.

The percentage contribution of these main sectors to the GNP are as follows:

- agriculture   13%
- industry   25%
- construction   5%
- services   55%
- other sectors   2%

Amongst services, transportation constitutes some 15% and shipping is placed within this category.

Shipping business is not limited within the boundaries of Turkey and therefore, it is planned for Turkish shipping to increase its efficiency within world trade.

The Role of Turkish Shipping in Foreign Trade

The role of Turkish shipping in the Turkish economy is illustrated in Figure 12 for the period between 1990-1994. The categories of total Turkish foreign trade by sector are for 1995:-

- general cargo   39.20%
- wheat and grain   3.80%
- mineral ore   7.70%
- coal   8.80%
- oil   40.50%

The Turkish total carriage of imports by sea between 1985-1996 is illustrated in Figure 13. It is clear that the total imports carried by sea has more than doubled in this period from 29.7 million tons in 1985 to 72.8 million tons in 1996.

Similarly, the amount of Turkish total carriage of exports by sea during the same period is illustrated in Figure 14. Again, similar to the carriage of the imports by sea, the amount of export goods carried by sea during the same period has almost doubled. The amount of exports in 1985 was 11.5 million tons and became 18.9 million tons in 1996.

The total amount of goods carried by sea for import and export increased from 41.1 million tons in 1985 to 91.2 million tons in 1996.

Figure 12: Turkish Foreign and Cabotage Carriage

Year	A	B	C	D	E	F
1990	53.1	43.9	15.2	22.2	37.7	48.9
1991	70.2	49.9	20.3	22.7	32.3	41.0
1992	72.4	50.5	21.9	29.5	40.8	48.6
1993	83.0	64.9	18.1	33.5	40.4	54.4
1994	74.7	52.6	22.1	37.0	49.5	71.0

Source Turkish Lloyd, 1995.

A -   Carriage of goods by sea (million tons)
B -   Imports (million tons)
C -   Exports (million tons)
D -   Carriage of foreign trade by Turkish vessels (million tons)
E -   % of Turkish vessels in carriage of foreign trade
F -   Turkish vessels total foreign trade and cabotage (million tons)

Conclusions

The capacity of the Turkish merchant fleet was 4.5 million dwt in 1985 and had increased to 10.6 million dwt at the end of 1997. This in itself can be considered a success in the development of Turkish shipping.

There have been no state subsidies for the shipping industry during the period 1984-1998 and shipowners have been financing the industry without any financial support from the government. Meanwhile, many other industries in Turkey have continued to receive subsidies from the state for investment. The shipping industry in Turkey has continued to develop its economic role without further artificial support.

Despite this success, the Turkish shipping industry needs subsidy from the government for further investment and if it is to play a full role in the future particularly in the context of liberalisation of the economy in Turkey.

The shipbuilding sector has had an important effect in the development of the Turkish shipping industry, which this paper has not attempted to analyse. The shipyards are operating with a considerable amount of overcapacity at present in Turkey. The increase in the Turkish merchant fleet has been mostly through the purchase of second-hand vessels. This dominance of second-hand purchases should not be the case and more importance needs to be given to the new building of vessels to overcome two problems. The first stems from the continuing over-capacity of the shipbuilding industry; and the second from the ageing characteristics of the Turkish fleet with an increasing predominance of old, second-hand vessels.

Figure 13: Carriage of Turkish Imports by Sea 1985-1996 (million tons)

Year	Turkish vessels	Foreign vessels	Total import
1985	15.8	13.9	29.7
1986	13.0	16.0	29.0
1987	16.7	18.9	35.6
1988	14.7	18.1	32.8
1989	15.4	18.3	33.7
1990	17.6	26.3	43.9
1991	16.7	33.2	49.9
1992	21.3	29.3	50.6
1993	24.9	40.0	64.9
1994	26.5	26.1	52.6
1995	27.2	36.8	64.0
1996	28.6	44.2	72.8

Source: Chamber of Shipping, 1998.

Figure 14: Carriage of Turkish Exports by Sea 1985-1996 (million tons)

Year	Turkish vessels	Foreign vessels	Total export
1985	2.5	9.0	11.5
1986	4.7	8.7	13.4
1987	4.4	8.6	13.0
1988	5.0	14.7	19.7
1989	5.2	16.3	21.5
1990	4.7	10.5	15.2
1991	6.0	14.3	20.3
1992	8.3	13.6	21.9
1993	8.6	9.5	18.1
1994	10.5	11.6	22.1
1995	8.0	12.2	20.2
1996	7.4	11.4	18.8

Source: Chamber of Shipping, 1998.

Meanwhile, various state-owned maritime institutions are on the list for early privatisation including a number of ports and shipyards.

Additionally, Turkish Cargo Lines, which is one the biggest operators of general cargo vessels and bulk carriers, is also included in that list.

The shipping industry continues to bring in over 5 billion US$ a year to the Turkish economy and thus will continue to be one of the locomotive industries of the economy. The merchant fleet approaches a capacity of 12 million dwt, with extensive shipyards and international ports which will be made more productive by privatisation, and with a greater political role promised by successive administrations.

The main problem Turkey now faces is with the economy on a macroscale and in particular with its instability and uncertainty resulting in a series of crises. This has the effect of interrupting domestic economic demand with a knock-on effect for the shipping industry. The fact that shipping is an international activity, does help to ameliorate the effects but does not remove them altogether and a greater involvement of the state to overcome these periodic difficulties is needed if the industry is to make the greatest returns to the Turkish community.

References

Caglar, E. (1995) *Congress of Turkish Maritime in the Future*, Izmir, Turkey.

Chamber of Shipping (1995) *Sector Report of Turkish Shipping 1994*, Istanbul, Turkey, pp. 163, 164, 173, 182, 183, 223, 336, 337.

Chamber of Shipping (1995) *Turkish Shipping World*, October, Istanbul, Turkey, p.42.

Chamber of Shipping (1996) *Turkish Shipping World*, January, Istanbul, Turkey, pp. 15, 18, 20, 45.

Chamber of Shipping (1996) *Turkish Shipping World*, February, Istanbul, Turkey.

Chamber of Shipping (1996) *Turkish Shipping World*, March, Istanbul, Turkey, p.35.

Chamber of Shipping (1998) *Report on Shipping Sector - 1997*, Istanbul, Turkey, pp. 69, 93-95, 99-101.

Drewry Shipping Consultants (1995) *The Shipbuilding Market*, London, p. 37.

Interview with Mr. Cengiz Kaptanoglu (04.04.1996) Istanbul, Turkey.

Interview with Mr. Adil Goksu (03.04.1996) Istanbul, Turkey.

Lloyd's List (1993) *Turkey*, October 25th, Lloyd's of London Press, London.

Lloyd's List (2000) *Turkey to Streamline Maritime Affairs*, October 3rd, Lloyd's of London Press, London.

Lloyd's Ship Manager (1993) *Turkish Shipping and Ports Directory 1993*, August, Lloyd's of London Press, London, pp. 11, 13, 17.

Lloyd's Ship Manager (1994) *Turkish Shipping and Ports Directory 1994*, Lloyd's of London Press, London, p.15.

Middle East Research Institute (1985) *Meri Report: Turkey*, University of Pennsylvania, Croom Helm Ltd., USA, p. 5.

Ministry of State (1992) *Towards the Ministry of Maritime*, Ankara, Turkey, p. 15.

Seatrade (1993) November, p. 113.

Seatrade (1995) July, p. 61.

State Institute of Statistics (1995) *Statistics and Analysis of Turkish Economy*, Ankara, Turkey, 1, 7.

Tez, T. (1993) *What We Have Mentioned for Shipping*, Ankara, Turkey, p. 13.

Turkish Maritime Organisation (1989) *Sector Report of Ship Operators*, Istanbul, Turkey, p. 6.

Turkish Naval Force (1996) *Proceedings of the Conference on the Problems of Turkish Shipping*, Kocaeli, Turkey, pp. 43-48.

Yercan, F. and Roe, M. (1999) *Shipping in Turkey*, Ashgate: Aldershot.

3 Dimensions of Service Quality in Container Transportation: an Empirical Investigation

OKAN TUNA
SCHOOL OF MARITIME BUSINESS AND MANAGEMENT,
DOKUZ EYLUL UNIVERSITY, IZMIR

Introduction

From the point of view of competitive market conditions, it is indeed inevitable that those companies providing container transportation, supplying services for the business markets and thus affecting the competitive positions of shippers within foreign markets, have to be customer oriented. As a consequence, these companies dealing with container transportation have got to enhance and accelerate their attempts at improving their service quality in compliance with the steadily increasing shipper expectations that continue to emerge.

This study aims to determine the service quality perception of Turkish shippers in terms of "expectations" and "service performance" and to investigate the dimensions of service quality in the case of a business-to-business service.

Service Quality

Service quality has been described as a form of attitude, related but not equivalent to satisfaction, resulting from the comparison of expectations with performance (Bitner, 1990). In this sense, quality is what customers perceive. Perceived service quality differs from satisfaction in that service quality refers to the customer's attitude or global judgement of service superiority over time, while satisfaction is considered to be connected with a specific transaction (Parasuraman *et al*, 1985).

The quality of a service, as customers perceive it, has two dimensions;

technical and functional (Groonroos, 1990). The technical dimension of quality is related to *what* the customers receive in their interactions with the firm and mainly covers the activities within the technical core of the service process. On the other hand, the functional dimension of quality is related to *how* the customer receives the service and covers the contact personnel and physical facilities. Groonroos also has stated the importance of corporate and/or local image which can impact upon the perception of quality in various ways.

Parasuraman et al (1985) developed five different determinants within service quality after a series of qualitative and quantitative studies; tangibles, reliability, responsiveness, assurance and empathy. The Service Quality Model (SERVQUAL) developed by Parasuraman *et al* considers the difference between the expectations of customers and service performance in order to measure service quality. On the other hand, Cronin (1992) and Taylor (1994) have criticised SERVQUAL and proposed that service performance (perception) alone is adequate in measuring service quality (SERVPERF).

Container Transport Services and Service Quality

Container transport companies serve organisational markets and not a customer but the buying centre plays an important role in performance evaluation. Considering this fact, measurement of service quality within container transport services is relatively different compared with consumer markets. Container transport services can be regarded as the "possession processing" services in which there is no real need for customers to enter the service factory and accompany their possession while it is being processed (Lovelock, 1996); whereas, a "medium-contact" occurs between the shipper and container transport service provider during the service delivery.

Considering the importance of attributes used in mode/carrier selection criteria research, this literature is reviewed as well as that relating to container transport service quality. Bardi et al (1987) found five factors important to the shippers in carrier selection; transit time reliability, transportation rates, total transit time, willingness to negotiate, and financial stability. McGinnis (1979) has found that on time pick up, delivery, reliability, and transit time are of critical importance to traffic managers. Matear and Gray (1993) showed that fast response to problems, avoidance of loss or damage and on-time collection and delivery are the most important service attributes for shippers.

D'este and Meyrick (1992) targeted the shippers of ro/ro ferries in order to evaluate the relative importance of selection attributes. Rapid and reliable cargo movement, price, and fast response to problems were found to be the most important attributes. Pearson (1981) identified flexibility, frequency of sailing, transit time, reliability, and regularity as the quality dimensions in container line services in terms of selection criteria. Another study (Kent and Parker, 1999) on containership selection criteria revealed that reliability, equipment availability and service frequency are the most important factors for shippers. Brooks (1995) sampled shippers including both freight forwarders and consignees in seven countries and found that reliability and responsiveness are the most critical determinants of service quality.

Methodology

A private Turkish line was sponsor of the study. However, this study was restricted only to the users of the Port of Izmir. As far as both the demand and supply sides are concerned, the Port of Izmir has an ideal location for this research. On the supply side, Izmir is the largest port in Turkey with a 388,172 TEU container handling capacity. On the demand side, Izmir is an important trade centre in Turkey with its hinterland.

Two 34 item rating scales were developed in terms of both "expectations" and "service performance" (Tables 1 and 2). On the other hand, a nine item rating scale was developed for the behavioural intentions of shippers. A five point Likert scale was used with anchors at 1 (strongly disagree) and 5 (strongly agree).

Questionnaires were mailed with a cover letter to 400 customers of the Port of Izmir liner agency in 1999. All customers were reminded 20 days after postage. A 21.75% (87 shippers) response rate was achieved.

Of those 87 respondents, 82.8% were exporters and only 17.2% had been operating as freight forwarders. Some 48.3% of the respondents had experience of 10 years or more in container transport buying and 63.9% of the exporter respondents operated in the food/chemical industry.

Expectations and Service Performance

Issuing accurate bill of lading and delivering the cargo without damage were determined as the most important factors within expectations (mean = 5.00). Other important factors were issuing accurate price quotations, responding to complaints quickly and providing clean and undamaged containers (mean = 4.98) (Table 1).

As far as the service performance of the liner company was considered, polite and respectful personnel has the highest mean (4.73). On the other hand, issuing accurate invoices (mean = 4.69) and informing correctly about transhipment (mean = 4.67) were ranked as the other important factors within service performance (Table 2).

Table 1: Top Ten Findings in Terms of Expectations

VARIABLES	Rank	Mean	Standard Deviation
Issuing accurate bill of lading	1	5.00	0.00
Delivering the cargo without damage	2	5.00	0.00
Issuing accurate price quotations	3	4.98	0.15
Responding to complaints quickly	4	4.98	0.15
Providing clean and undamaged containers	5	4.98	0.15
Expert and knowledgeable personnel	6	4.95	0.26
Dependability in handling problems	7	4.95	0.21
Issuing accurate invoices	8	4.95	0.21
Responding to urgent deliveries quickly	9	4.94	0.23
Issuing invoices on time	10	4.93	0.25

Reliability and Validity of the Scale

The Cronbach Alpha coefficient (α) has a value of 0.84 which indicates the reliability (high internal consistency) of the service quality scale. On the other hand, the convergent validity of the scale is significant (Pearson correlation coefficient = 0.526; $p = 0.000$).

Table 2: Top Ten Findings in Terms of Service Performance

VARIABLES	Rank	Mean	Standard Deviation
Polite and respectful personnel	1	4.73	0.70
Issuing accurate invoices	2	4.69	0.70
Informing correctly about transhipment	3	4.67	0.85
Providing available containers	4	4.65	0.59
Willingness of personnel to help	5	4.62	0.73
Convenient working hours for contact	6	4.60	0.82
Dependability in handling problems	7	4.56	0.77
Issuing accurate price quotations	8	4.54	0.81
Giving clear and correct information about costs	9	4.53	1.00
Delivering the cargo without damage	10	4.53	0.83

Dimensions of Service Quality in Container Transportation: Factor Analysis

Principal components analysis with Varimax rotation was employed in order to identify the major dimensions of the container transportation service quality. Table 3 reveals the dimensions derived - reliability, personnel service, accurate and on-time documentation, responsiveness, competence, supporting activities, integrated service, security, information, value added service, communication and export knowledge.

These 12 factors accounted for 73.74% of the total variance. Reliability as the first factor accounted for 10.76% of the total variance while personal service as the second factor accounted for 8.01% of the total variance.

Table 3: Container Transport Service Quality: Factor Analysis (Pt. 1)

FACTORS	α	1	2	3	4
Factor 1: Reliability	**0.8260**				
Issuing accurate bill of lading		0.744			
Providing clean and undamaged containers		0.669			
Delivering cargo without damage		0.617			
Issuing bill of lading quickly		0.600			
Delivering cargo at promised time		0.592			
Giving clear and correct information about costs		0.563			
Informing correctly about transshipment		0.556			
Factor 2: Personal Service	**0.7722**				
Polite and respectful personnel			0.848		
Willingness of personnel to help			0.847		
Expert/knowledgeable personnel			0.530		
Factor 3: Accurate and On-time Documentation	**0.6467**				
Issuing accurate price quotations				0.803	
Issuing accurate invoices				0.682	
Issuing invoices on time				0.657	
Responding to urgent deliveries quickly				0.472	
Factor 4: Responsiveness	**0.5931**				
Unloading the container at the arrival port					0.746
Responding to enquiries promptly					0 718
Responding to complaints quickly					0.471
Eigenvalue		7.423	2.281	2.225	2.120
Total Variance (%)		10.766	8.018	7.365	6.526

Table 3: Container Transport Service Quality: Factor Analysis (Pt. 2)

FACTORS	α	5	6	7	8	9	10	11	12
Factor 5: Competence	**0.6454**								
Having modern looking equipment		0.796							
Helping company personnel during stuffing		0.576							

Table 3: Container Transport Service Quality: Factor Analysis (Pt.2 cont.)

Using information technology		0.527				
Providing available containers		0.457				
Factor 6: Supporting Activities	**0.6206**					
Informing about condition of cargo			0.819			
Informing about technical details			0.731			
Convenient working hours for contact			0.521			
Factor 7: Integrated Service	**0.4534**					
Providing protective packaging				0.664		
Dependability in handling problems				0.593		
Factor 8: Security						
Keeping company secrets					0.797	
Factor 9: Information	**0.4108**					
Informing about customs related issues						0.839
Informing about ship and other equipment						0.417

Table 3: Container Transport Service Quality: Factor Analysis (Pt.2 cont.)

Factor 10: Value Added Service	0.3528							
Minimum changes to schedules						0.859		
Informing on the method of stuffing						0.404		
Factor 11: Communication	0.3235							
Informing of changes to schedules							0.725	
Giving arrival notices on time							0.667	
Factor 12 : Export Knowledge	-							
Informing about export issues								0.735
Eigenvalue	1 .817	1.627	1.447	1.374	1. 330	1 .258	1.163	1.014
Total Variance (%)	6.313	6.284	5.743	4.807	4.699	4.468	4.385	4.369

Conclusions

This research has investigated the service quality perceptions of Turkish shippers and attempted to find out the dimensions of service quality within business-to-business service. A valid and reliable service quality scale has been constructed in order to measure service quality in container transport services. As far as the managerial implications are considered, reliability seems to be the most important factor for marketers in order to increase the loyalty of their customers. On the other hand, they need to consider personnel service in establishing service delivery system of the container transport companies.

As already stated, this scale has been applied to the shipper using the Port of İzmir, and there might be a need to apply this service quality scale to different ports and different types of shippers. Another important point is to consider the logistical services of container transport companies while applying such a quality scale.

References

Anon (1994) SERVPERF vs SERVQUAL: Reconciling Performances Based Perception - Minus - Expectations Measurement of Service Quality, *Journal of Marketing*, 58, 125.

Bardi, E.J., Bagchi, P.K. and Raghunathan, T.S. (1987) Motor Carrier Selection in a Deregulated Environment, *The Logistics and Transportation Review*, 23, 4, 5.

Bitner, M.J. (1990) Evaluating Service Encounters: The Effects of Physical Surroundings and Employee Responses, *Journal of Marketing*, 54, 69.

Brooks, M.R. (1995) Understanding the Ocean Container Market - A Seven Country Study, *Maritime Policy and Management*, 22, 1, 39-49.

Cronin, J.J. and Taylor, S.A. (1992) Measuring Service Quality: A Re-examination and Extension, *Journal of Marketing*, 56, 55.

D'Este, G.M. and Meyrick, S. (1992) Carrier Selection in a Ro/Ro Ferry Trade Part 1: Decision Factors and Attitudes, *Maritime Policy and Management*, 19, 2, 115.

Groonroos, C. (1990) *Service Management and Marketing: Managing the Moments of Truth in Service Competition*, Lexington Books, Toronto, p. 37.

Kent, J.L. and Parker R.S. (1999) International Containership Carrier Selection Criteria: Shippers/Carriers Differences, *International Journal of Physical Distribution and Management*, 29, 6, 40.

Lovelock, H.C. (1996) *Services Marketing*, Prentice Hall, U.S.A, p. 30.

Matear, S. and Gray, R. (1993) Factors Influencing Freight Service Choice for Shippers and Freight Suppliers, *International Journal of Physical Distribution and Logistics Management*, 23, 2, 25.

McGinnis, M.A. (1979) A Comparative Evaluation of Freight Choice Models, *Transportation Journal*, 29, 2, 36.

Pearson, R. (1981) *Containerline Performance and Service Quality*, Marine Transport Centre, University of Liverpool, pp. 120-121.

Parasuraman, A., Zeithaml, V.A. and Berry, L.L. (1985) A Conceptual Model of Service Quality and its Implications for Future Research, *Journal of Marketing*, 49, 41.

Parasuraman, A., Zeithaml, V. and Berry, L. (1988) SERVQUAL: A Multiple Item Scale for Measuring Consumer Perceptions of Service Quality, *Journal of Retailing*, 64, 12.

Zeithaml, V.A., Berry, L.L. and Parasuraman A. (1996) The Behavioural Consequences of Service Quality, *Journal of Marketing*, 60, 33.

4 Regional Development and Port Planning: The Port of Mersin as a Feeder Terminal in the Eastern Mediterranean[1]

GÜLER BILEN ALKAN AND MAHMUT CELAL BARLA
DEPARTMENT OF MARITIME TRANSPORTATION BUSINESS,
ISTANBUL UNIVERSITY
A. GÜLDEM CERIT
DOKUZ EYLUL UNIVERSITY, IZMIR

Introduction

The increase in world container traffic has affected ports in the Eastern Mediterranean and in particular the Port of Mersin on the Southeast Mediterranean coast of Turkey has become strategically significant. Mersin is benefiting from the increased maritime traffic in the region, Turkey's increasing involvement with international markets, and developments in South-eastern Anatolia.

Starting operations in the year 1962 and located at 36 degrees 47' 30" north and 34 degrees 38' 00" east, the Port of Mersin is the largest port in Turkey with respect to its berth length of 3,292 m. and port area of 994,000 square meters. With respect to total loadings/unloadings, in 1999 Mersin stood on the first row amongst Turkish ports, with an annual figure of 13,392,574 tons, with container handling of 244,002 TEU during the same period.

Due to the growth potential of seaborne trade in the region, Turkey's leading free trade zone is also situated at Mersin. Currently Mersin Free Zone maintains about half of the total trade volume generated by the total seven free trade zones of Turkey.

The most important impact of the economic environment surrounding the Port of Mersin is the outstanding growth of the South-eastern Anatolia

[1] This paper was originally presented at the IAME Halifax Conference in September 1999.

region. With the implementation of the South-eastern Anatolia Project (Guneydogu Anadolu Projesi - GAP) the dams that are planned on the Euphrates and Tigris rivers will help irrigate more than 1.7 million hectares of land and 27 billion kW of electric energy will annually be produced. Obtaining 2-3 crops each year under irrigated conditions has begun, new crops like soya-beans have been introduced and production of some crops are expected to increase appreciably. By the early 2000s, cotton production in the region is expected to exceed what the whole country produces today. Turkey is the world's seventh largest cotton producer and the sixth largest textile/apparel exporter, the textile industry being the largest exporter among all industrial sectors in Turkey. With the completion of the South-eastern Anatolia Project (GAP), cotton production and the corresponding exports are expected to increase considerably, accelerating maritime trade in the region.

With these factors and estimates at hand, the Port of Mersin is to be organised as a feeder container terminal in the Eastern Mediterranean and this study outlines an approach to the strategic planning of the port within this context.

Regional Development and Port Services

Like an airport or a railway station, a seaport is a point where different transport modes intersect, and transfers from one to another take place. The degree of need for port services depends on trends and events in other transport modes like roads and railways as well as shipping. The international nature of a port also means that decisions about its capacity and shape are influenced by distant events - trends in world trade, decisions by foreign shippers and shipbuilders, decisions by domestic exporters and the trends in regional development projects in the country.

Ports are part of a larger transport system. If as a result of some improvement in the port, goods are able to move through more quickly, the transport system as a whole could benefit from more rapid throughput and lower costs. Improvements at one port can affect the movement of traffic through other ports in the country or those of neighbouring states.

In recent years, ports have had to accommodate far reaching technological changes. The pace of these technological changes is being set by the economic circumstances of the developed countries. The smaller and poorer developing countries have to adapt their port facilities to the type of ship being sent, or risk becoming backwaters off the main shipping lanes. One alternative to expansion to cope with the largest container ships is to plan to be a feeder port linked to a larger port that is directly served by larger container ships.

A seaport in its function as a traffic link between the manufacturer on the one hand and the commercial markets on the other hand, has to assure speedy handling. Ports are vital points, holding more or less a key function for economic development.

Planning a port aims to fulfil all port functions and for planning purposes political, legal, economical, sociological, technical, physical and natural objectives are concerned. The weight of different factors will very much depend on the status of the port and its form of organisation.

Planning is a methodological tool for solving problems. But planning itself does not solve problems, it has to take into account a given potential of resources, which may be and normally are limited. Normal conditions presumed, an enlargement or replacement of facilities is needed. Economic and natural resources and development projects to enlarge these resources affect port planning directly. Port development effects are discussed in Figure 1. As may be noted, regional development effects have an impact on both economic and social development effects and the reverse is also valid, that is, regional development is also affected by economic and social development issues. Furthermore, as an overall result, regional development accelerates development of trade activities and the increasing need for logistics and transportation services reflects the importance of port development effects.

Economic effects of regional development may be either direct, or regional or national. Social development effects on the other hand may be grouped as community effects and national land development effects. These in turn interactively affect trade creating effects and port development effects.

The effective management of agricultural activities is one of the critical components of regional development and rural change. The principle components of an agricultural project are the natural resources base, the economic and social structure, agricultural inputs, and the institutions that will hold it together. These component parts will interact. None of them in practice is wholly changeable, but all can be modified and improved (Bridger and Winpenny, 1996). Both a port and regional development have socio-economic development effects on the regional community and the nation at large through its various functions.

Port Development Effects

↑↓

Economic Development   Social Development Effects
Effects

↑↓   ↑↓

Direct Economic Effects   Community Effects
Regional Economic Effects   National Land Development Effects
                                      National Economic Effects

↑↓

Regional Development Effects

Figure 1: Interaction of Regional and Port Development Effects
Source: Haruo Okada, Port Development in Japan, Overseas Coastal Area Port Development Institute of Japan, Unpublished special lecture notes, Tokyo, 1988, pp.1-6.

Container Ports and Developments in the Eastern Mediterranean

The process of concentration in shipping - larger ships, more transshipment, mergers, alliances - has profound effects on port development. The number of port calls by post-Panamax vessels will be reduced as long as the additional costs for feeder and intermodal connections are lower than the savings from fewer port calls. This tendency directly leads to further concentration of port traffic in fewer and larger ports. The rapid development of transshipment has radically changed the market with significant changes in the main ports (Hoffmann, 1998).

Hub ports are known as those ports concentrating on the container flows of export, import and also transshipment activities. Feeder ports on the other hand mainly fulfil the requirements of feeder vessels. As a result of the concentration of containers, liner shipping companies are able to realise scale economies and reduce operational costs at terminals.

Essential Requirements for Hub Ports and Feeder Ports

Hub ports are container transshipment ports for mother vessels. They tend to reduce the number of calls and choose the most strategic locations as transhipment centres, with advantages in terms of service requirements and of economics of modern operations (IAPH, 1999). Essential factors for the success of hub ports may be summarised under the following headings:

- economic and political stability,
- friendly relationships with neighbour countries,
- feeder service network,
- completed inland transportation infrastructure,
- cheaper terminal costs,
- simple procedures in the terminal,
- free use of terminal/possibility to maintain ship schedule,
- efficiency of operational software, EDI, etc.

For the success of port development, the port authority or governments have to choose their policies for the development of hub or feeder ports. If development policy leads to the choice of a hub port, then the infrastructure covering the facilities and the software to attract ships have to be introduced accordingly. Especially heavy investments are required to establish hardware. The requirement for 15-meter depth and supergantry cranes to serve the over Panamax containerships are some examples of these investments (Chadwin, *et al,* 1990).

Container feeder ports usually include some of the following advantages for the customer that are designed mainly to attract container feeder vessels:

For transit containers:

- discount rates for discharging/loading and other cargo handling charges,
- container stockyard.

For vessels discharging and loading transit containers:

- berthing priority,
- 24 hours daily working,
- discount rates for ships' handling charges.

Container Movement in the Mediterranean

World container throughput in 1997 was estimated to be 170 million TEUs (IAPH, 1999). It is estimated that the Mediterranean accounts for over 10% of this. Total container traffic in the Mediterranean grew to 9.9 million TEUs in 1994 from 4.7 million TEUs in 1984, the annual growth rate being around 7.2%, a little lower than the world rate of 9.01% (Containerisation International, 1998). On the other hand transhipment movements in the area have risen from 1.69 million TEUs in 1992 to 2.342 million TEUs in 1996 (IAPH, 1999).

Sixteen countries in total surround the Mediterranean Sea and these countries each have main container handling ports, except Albania. There are 40 top ports handling containers in the Mediterranean region. Using 1997 figures, the port with the largest handling volume is Algeciras in Spain, followed by La Spezia in Italy and Barcelona in Spain. Piraeus and Limassol are the major container feeder ports in the east Mediterranean area, but Port Said, Alexandria, Larnaca, and Valleta have experienced increased transshipment container handling.

Massive traffic increases at transshipment hubs such as Gioia Tauro are matched by huge expansion plans for new facilities at Cagliari, Taranto, Sines, Tanger-Atlantique, Gibraltar and East Port Said, with some of these projects more advanced than others (Drewry, 2000). There is fierce competition between the major hub ports in the region and furthermore there will be increased competition from newcomers.

The trend of container movement in the Mediterranean region can be analysed by dividing the region into two areas, such as the east and west Mediterranean. The east Mediterranean handled 3.7 million TEUs in 1997, 62% less than that handled by the west (9.9 million TEUs). However, since 1987, container movement in the east Mediterranean has been rapidly increasing with an annual growth rate of 10-20% (Containerisation International, 1998).

Container throughputs in the Mediterranean ports are given in Table 1. The larger Mediterranean ports reported a total of 10 million TEU handled in 1997 against 3 million TEU in 1987.

In 1996, the country handling the greatest volume of containers in the Mediterranean was Italy (3.7 million TEUs), followed by Spain (3.5 million TEUs) and France (1.8 million TEUs). However it should be noted also that France faces the North Sea. Israel handled 959,824 TEUs followed by Greece, Egypt, and Turkey as the second most significant group. Egypt had handled around 180,000 TEUs/year or so until 1989 when Damietta was not fully operational, but since 1990 the traffic has remarkably increased and in 1992 Egypt ranked first in the east and third overall. Turkey generally lies fourth in the east (Containerisation International, 1998).

Table 1: Container Traffic in the Mediterranean Ports (000 TEU)

Port	1987	1995	1997
Algeciras	430	1,155	1,538
Gioia Taura	-	-	1,488
Genoa	265	615	1,180
Barcelona	385	689	972
Valencia	321	672	832
Marsaxlokk	-	-	663
Marseilles/Fos	388	490	622
Piraeus	266	600	617
La Spezia	199	965	616
Damietta	-	450	600
Leghorn	478	400	501
Limassol	141	294	399
Izmir	-	252	388
Mersin	-	146	234
Larnaca	102		165
Haifa	-	460	-
Port Said	-	289	-
Ashdod	-	276	-

Source: Dynamar-Dyna Liner, Weekly Magazine, 34/1998, 21 August

Container Movement by Port

Table 2 analyses the container movement by ports in the Mediterranean. The characteristics of the ports are also stated in the table. Major ports are regular call ports by mother vessels, whereas hub ports are container transshipment ports and feeder ports are the feeder vessel calling ports. Major ports handle larger quantities of containers and belong to developed countries like Italy, France and Spain that have large levels of foreign trade inside and outside the Mediterranean Sea. The eastern Mediterranean ports specified in Table 2 serve south Europe, North Africa and the Middle East and account for half of the top 40 ports.

In the east Mediterranean, Damietta, which handled no containers before 1985, became the largest port for container handling in 1994. From 1990, cargo volume has dramatically increased because the port of Damietta has begun playing the function of hub-port in the east Mediterranean. This has affected container handling in Piraeus (Greece), Limassol (Cyprus) and neighbouring Alexandria which formally played central roles.

Concerning ports in Turkey, Izmir is the largest port for container handling and cargo volume has been increasing with a 21.5% annual growth rate starting from 1990. However, the volumes for the second largest port Haydarpasa and the third largest Mersin have been nearly constant. The cargo volume of Haydarpasa has dramatically decreased in 1994 (Containerisation International, 1998).

Service Routes, Feeder Service and the Service Network in the Mediterranean

Major container service routes to/from the Mediterranean Sea area, are as follows:

- UK and Northern Continental Europe,
- Southern Europe (Mediterranean Sea),
- USA,
- Far East.

The range of prevailing container ship sizes is very wide, starting from 400 TEU to even 8,000 TEU types. As for Southern European (Mediterranean Sea) services, these are entirely within the Mediterranean region. Smaller vessels up to the 500 TEU type, are in operation, similar to the feeder services in this area.

Due to the large volume of maritime trade in the Mediterranean, container feeder services prevail in the area to serve the many container ports situated along the Mediterranean coast. Container feeder services are the result of container ship operators' needs to reduce ships' operating costs and to shorten the length of voyages. This can be accomplished by minimising the number of regular calling ports without however, reducing freight earnings in view of the progress of container ship enlargement and consequently sizeable ship construction cost as well as crew costs. This trend has been proceeding in other container trade services since the sizes of container ships reached more than 3,000 TEU capacity.

Table 2: Main Container Ports in the Mediterranean

Port	Country	Feeder Ports (F. P.)	Hub Port (H.P.)	Major Ports (M. P.)	Line
Algiers	Algeria	F.P.			NA and ME[2]
Larnaca*	Cyprus		H.P.		NA and ME[2]
Limassol*	Cyprus		H.P.		NA and ME[2]
Alexandria*	Egypt		H.P.		NA and ME[2]
Damietta*	Egypt		H.P.		NA and ME[2]
Port Said*	Egypt		H.P.		NA and ME[2]
Marseilles	France			M.P.	SE[1]
Piraeus*	Greece		H.P.		NA and ME[2]
Haifa	Israel	F.P.			NA and ME[2]
Genoa	Italy			M.P.	SE[1]
Gioia Taura	Italy		H.P.		SE[1]
La Spezia	Italy			M.P.	SE[1]
Leghorn	Italy			M.P.	SE[1]
Naples	Italy			M.P.	SE[1]
Ravenna*	Italy			M.P.	SE[1]
Trieste*	Italy			M.P.	SE[1]
Beirut*	Lebanon	F.P.			NA and ME[2]
Benghazi	Libya	F.P.			NA and ME[2]
Tripoli	Libya	F.P.			NA and ME[2]
Valletta	Malta		H.P.		NA and ME[2]
Algeciras	Spain			M.P.	SE[1]
Barcelona	Spain			M.P.	SE[1]
Valencia	Spain			M.P.	SE[1]
Lattakia*	Syria	F.P.			NA and ME[2]
Tunis	Tunisia	F.P.			NA and ME[2]
Mersin*	Turkey	F.P.			NA and ME[2]
Izmir*	Turkey	F.P.			NA and ME[2]
Istanbul*	Turkey	F.P.			NA and ME[2]

* East Mediterranean ports [1]South Europe [2]North Africa and Middle East
Source: Based on data from Containerisation International Year Book, 1998.

There are 28 container (main ship) service routes related to the Mediterranean Sea. Eleven routes are present between the Far East/Japan-Mediterranean, 5 routes Far East/Japan-Europe which stop or are bound for ports in the Mediterranean, and 3 Round the World routes which stop at some ports on the way (International Transportation Handbook, 1997).

Ports served by ships of these routes are divided into those of the east and west Mediterranean. Major calling ports in the east Mediterranean are Haifa and Alexandria, while in the west Mediterranean, Algeciras, Barcelona, Genoa, Valencia in Spain and Marseilles in France. In Turkey, Izmir and Mersin are served by feeder ships.

Southeast Anatolia Project and the Port of Mersin

The Port of Mersin on the Southeast Mediterranean coast of Turkey serves maritime traffic in the region under the conditions of Turkey's increasing involvement with international markets and the developments in South-eastern Anatolia. The main economic impact of the environment covering the Port of Mersin is the outstanding growth of the South-eastern Anatolia region. With the implementation of the South-eastern Anatolia Project (GAP), the dams that are planned to be built on the Euphrates and Tigris rivers will help irrigation of more than 1.7 million hectares of land thus ending in an agricultural boom in the region which will cause substantial growth in trade and related activities.

In this study, GAP is considered as a regional development project and the organisation of the Port of Mersin as a container feeder port as the related port development project. Both of these determinants have development effects over each other and thus in this study interrelationships between Port of Mersin and the GAP project are discussed from the point of view of port development issues.

Southeast Anatolia Project (GAP)

A foremost aim of Turkey is to eliminate interregional economic and social imbalances within its borders. The optimal use of land and water resources is an important means to achieve this aim. The most important investment scheme this country has undertaken in this endeavour is called the Southeast Anatolia Development Project. Its Turkish acronym is GAP (Guneydogu Anadolu Projesi).

There are briefly two reasons for the Turkish Government to design and implement such a large project in Southeast Anatolia (Tomanbay, 1998). First; Southeast Anatolia is endowed with good water and land resources, and Turkey aims to use these resources optimally for the sake of the whole region as well as Turkey. Therefore the Southeast Anatolia Project is being developed on the Euphrates and the Tigris Rivers and their branches which originate in Turkey. They supply the majority of Turkey's total surface flow and pass through Syria and Iraq to reach the Persian or Arabian Gulf. Second, Southeast Anatolia is the most backward region of Turkey. There are big economic and social differences between this region and the rest of Turkey. For instance, the per capita income in the region is only 47% of the per capita income of the remainder of the country. Besides that, many economic and social indicators such as electrical energy consumption per capita, the number of hospital beds per 10.000 people and manufacturing's share of the GAP region in Turkey's GNP make it clear that the region is the key to eliminating economic disparities with other parts of the country. The project on these two transboundary rivers aims to eradicate regional inequality, promote economic growth, and social stability in the region and ensure that the region becomes both developed and wealthy.

The GAP Project and Outcomes Regarding Regional Development

The GAP project area lies in south-eastern Turkey and covers nine provinces. This region is a part of Upper Mesopotamia, which is the cradle of ancient Mesopotamian civilisation. The total land area of the project corresponds to approximately 10% of Turkey's land area and the population of the project area accounts for about 9.5% of Turkey's total population according to recent figures. The project envisages the construction of 22 dams, 19 hydroelectric power plants and two irrigation tunnels on the Euphrates and Tigris Rivers and their tributaries.

The major elements of the project are the Atatürk Dam and Sanliurfa Tunnel systems. Both had been completed by 2000 and the ratio of irrigated land to total GAP area is expected to steadily increase from 2.9% to 22.8% whilst that for rain fed agriculture will decrease from 34.3% to 7.0% (http://www.dsi.gov.tr/gap.htm). On the other hand, 27 billion kWh of electricity will be generated annually on top of an established capacity of 7,460 megawatt. The area to be irrigated accounts for 19.0% of all the economically irrigatable area in Turkey (8.5 million hectares), and the annual electricity generation will account for 22.0% of the country's economically viable hydroelectric power potential (118 billion kW).

Moreover, some of the basic expected outcomes of the project are as follows: many Turkish crops will double or even triple in quantity. GAP will provide food self-sufficiency in Turkey and will create 3.3 million extra jobs. Turkey's national income will be 12.0% higher than it would have been otherwise and GRP (Gross Regional Product) will increase more than four times. Urbanisation will receive a boost in the region and rural migration will slow down considerably.

Table 3: Socio-economic Projection of GAP (1997 figures)

GRP (Gross Regional Product)	Billion TL		%
	1985	2005	2005
GRP	384,508	1,695,194	7.7
Agriculture	152,083	395,904	4.9
Industry	60,334	405,897	10.0
Construction	28,611	102,725	6.6
Services	143,480	804,121	9.0
GSBH/Per person	89,420	182,672	3.7

Source: Toyran, R. (1997).

Socio-economic Projection of GAP

An analysis of the increase in GRP (Gross Regional Product) in the GAP area is given in Table 3. As can be noted, it is estimated that GRP increase in industrial products will be the highest of all categories (Toyran, 1997). It is especially industrial sectors related to agricultural production that are expected to develop and will result in production and trade figures for those industries increasing.

Fresh Fruits and Vegetable Production in the GAP Region

Agricultural experts all over the world have agreed that the GAP region has sizeable agricultural development potential because of its extremely suitable climate, agricultural lands, low labour costs, big regional market and strategic position in relation to the Middle East and Former Soviet Union countries.

Turkey's agricultural production will be increased in the coming 20 years as a result of increased irrigated land. Whilst the production of fresh fruit and vegetables for the summer season will increase from 1.3 million tons to 4.5 million tons, those for the winter season will also increase from 27,691 tons to 608,441 tons (Tomanbay, 1998).

The aim of the project includes a feasibility study concerning management of the fresh fruits and vegetables after harvest in the South-eastern part of Turkey. Application of the project covers a pilot study related to storage, cooling and packaging of the fresh fruits and vegetables that will be produced in the region.

Export Oriented Strategies of the GAP Project

The strategy adopted in the GAP Master Plan for the region's development has the following four basic components (Tomanbay, 1998):

- To develop and manage soil and water resources for irrigation, industrial and urban uses in an efficient manner;
- To improve land use through optimal cropping patterns and better agricultural management;
- To promote manufacturing industries with emphasis on agro-related ones and those based on indigenous resources; and
- To provide better social services, education and employment opportunities to control migration and to attract qualified personnel to the area.

In short the GAP Master Plan's basic development scenario is to transform the region onto an export base for its agro-industrial products.

Maritime Transport Demand of the GAP

In South-eastern Anatolia, cotton production is expected to expand significantly as a result of the GAP. Currently about 50,000 hectares are irrigated as a direct result of the GAP, 80% of which is planted in cotton.

Agriculture and its related industries such as the textile mills are important to the Turkish economy since more than 40% of the population lives in rural areas and earn the bulk of their income from farming and related activities. As a result, agriculture and rural development are top priorities for the government. The massive investments in GAP represent the Turkish government's main commitment to agricultural development.

With the development of production in the region, increasing volumes of products in the GAP region will be transported using the facilities explained below:

- Transportation facilities within the GAP area;
- Transportation facilities for neighbouring areas (maritime transport is not possible or uneconomical);
- Transportation facilities to the Middle East countries (Iran, Iraq, Syria);
- Transportation facilities to the west coast of Turkey (maritime transport to Izmir, Istanbul, Bursa, Zonguldak and neighbouring cities);
- European Countries (both maritime and motorway transport);
- Others.

It is probable that the main containerisable cargoes from the GAP region are as follows: cotton, cotton textile, vegetables, fruit, steel furniture, petro-chemical products.

The Port of Mersin and its Role in the Eastern Mediterranean

The container throughput of Turkey has increased rapidly in recent years reaching 972,307 TEUs in 1998 from 588,341 TEUs in 1994 and 70,000 TEUs in 1987 (TCS, 1999). In 1998 the Port of Izmir transhipped 398,619 TEUs, whereas traffic at the Port of Mersin reached 241,865 TEUs (TCS, 1999).

The City of Mersin, landlord of Mersin port, is one of the trade centres of Turkey not only with its rich history, culture and tourism but also with its hinterland of vast and fertile agricultural areas. The port of Mersin can be counted amongst the ten largest ports in Europe and is visited by about 5,000 ships a year. As such, it is the largest Turkish port.

The port is readily accessible with a hinterland stretching from West Europe to the Middle East and the Asian countries and from Russia to the North African countries. The port's rail link and its easy access to international highways make it an ideal transit port for trade to the Middle East. With its infrastructure and equipment, cargo handling facilities, storage areas and its proximity to the free trade zone, Mersin is one of the most important ports in the east Mediterranean. For Turkey, it is the main port for the region's industry and agriculture.

Mersin Free Zone, adjacent to the port, extends over an area of 800,000 sq.m. with quay length of 500m. and water depths of 8m. to 10m., and offers substantial advantages for both domestic and foreign companies.

The Port of Mersin and its Infrastructure

The container terminal has a depth of 14m. and this allows the port to be a major transshipment centre for the Mediterranean Sea together with a support area suitable for extending as well as its existing potential. The facilities at the port handle general cargo, containers, dry and liquid bulk and ro-ro services. The port is protected by two breakwaters of 2,800m. and 1,525m. in length. The depth at the entrance is 14m. and between 12m. and 14.5m. inside the breakwaters (Mersin Chamber of Shipping, 2000).

The equipment at the port includes five quayside container gantry cranes of 40 tons capacity, 17 shore cranes of 3 to 35 tons capacity, nine reach stackers of 40 tons capacity, 18 rubber tyred transtainers of 40 tons capacity, 18 mobile cranes 5 to 25 tons capacity, nine container forklifts of 10 to 42 tons capacity, 54 general cargo forklifts of two to five tons capacity, 38 tugmasters, one loader, 10 tractors, 43 trailers of 40 tons capacity, nine trailers of 15 tons capacity, ten trailers of 15 tons capacity, ten trailers of 20 tons capacity and two pneumatics of 50 tons/hour capacity. The floating craft comprise one floating crane of 60 tons capacity, one pilotboat, five tugboats and four mooring boats.

The storage facilities of the Port of Mersin include 589,230 sq.m. of open and 22,340 sq.m. of closed storage and two CSF of 10,464 sq.m. The annual storage capacity of these areas is 8,109,024 and 562,992 tons/year respectively. The annual holding capacity of the container storage yard is 203,376 TEU/year. Total annual handling capacity of the port is 6,564,000 tons, berth capacity being 11,100,000 tons.

Container handling operations at the terminal are carried out by five quayside container gantry cranes of 40-60 tons capacity and five container forklifts of 10-42 tons capacity. There are two container freight stations of 9,000 and 1,309 sq.m. Another service available at the terminal is the provision of reefer facilities for refrigerated containers.

For general cargo handling there are 11 berths with a total length of 1,572 m. and with alongside depths of 6 to 12m. The berths and yard behind are well equipped with modern and high performance cranes and handling equipment.

For bulk cargo handling there are three berths with a total length of 708m. and with alongside depths of 10 to 14.5m. Loading rate is 2,400 tons/hour, discharging rate being 1,200 ton/hour.

The port has ro-ro facilities and there is a rail ferry terminal at the port. The terminal has 253m. quay length and 12m. depth alongside. It has 10 km. manoeuvring lines behind the quays for marshalling wagons.

The ATAS refinery terminal is available for tankers with one finger pier capable of berthing two tankers at a time. The depth at the entrance channel and the pier is 14m. Maximum length is 299m. Ten hose connections provide access to a 26" pipeline at shoreline (Mersin Chamber of Shipping, 2000).

Mersin Free Trade Zone

Free zones are defined as special sites within the country but deemed to be outside of the customs border and they are the regions where valid regulations related to foreign trade and other financial and economic areas are not applicable, are partly applicable or new regulations are tested. Free zones are also the regions where a more convenient business climate is offered in order to increase trade volume and export for some industrial and commercial activities as compared to other parts of the country.

In general, all kind of activities can be performed in Turkish Free Zones such as manufacturing, storing, packing, general trading, banking and insurance. Investors are free to construct their own premises, whilst zones have also available office spaces, workshops, or warehouses on a rental basis with attractive terms. All fields of activities open to the Turkish private sector are also open to joint ventures with foreign companies.

There are ten free trade zones in Turkey, namely Mersin, Antalya, Aegean, Istanbul Ataturk Airport, Trabzon, Istanbul Leather, Eastern Anatolia, Mardin, Menemen and Samsun. Annual trade volume reached 5.5 billion US$ in 1997. Mersin Free Zone produced 32.5% of this figure with 1.793 billion US$, standing on the first row among all zones (http://www.igeme.org.tr/english/freezone/free.htm).

Adjacent to the Port of Mersin, the Mersin Free Zone was established on 776,000 sq. m. of public estate. Commercial activities have been conducted since the beginning of 1988 and the zone is operated by

MESBAS-Mersin Free Zone Operator Inc. (Mersin Chamber of Shipping, 1999).

There are 376 licensed firms in the zone of which 72 are foreign. 219 firms operate in the field of purchasing and selling whilst 29 are in the field of production, 49 storage, 50 leasing and others are in the field of banking, insurance and repair-maintenance (Table 4).

Table 4: Number of Licensed Companies in Mersin Free Trade Zone

	Investors	Tenants	Total
Domestic	283	214	497
Foreign	41	67	108
Total	324	281	605

Source: Mersin: Mersin Chamber of Shipping Publication, 1998.

In Table 5, trends in the development of trade at Mersin Free Zone are given. As can be noted after 1994, the figures have almost doubled and by the completion of the GAP project it is estimated that the figures will be around 5 billion US$.

Table 5: Annual Volume of Trade at Mersin Free Trade Zone

Years	Trade Volume, US$
1994	927,740.401
1995	1,400,037.552
1996	1,650,132.308
1997	1,792,600.000
1998	1,697,068.000

Source: Mersin Chamber of Shipping Publication, 1998.

Mersin as a Container Feeder Port after GAP

In the meantime, the Port of Mersin is active as a container feeder port in the Eastern Mediterranean in competition with ports in Cyprus or Greece. Additionally, the Port is acting as a container transit port to Iraq due to the following competitive advantages:

- geographical location,
- sufficient capacity, space and equipment to handle containers (though further investment may be required in the future),

- experience in handling large numbers of containers,
- existence of the free-trade zone.

Conclusion

After the completion of the Southeast Anatolia Project (GAP), the region will face a high growth rate in the field of especially irrigation, agriculture, agriculture based manufacturing industries and energy production. Covering an area of 74,000 sq. km, agricultural production, especially cotton and grains, will be doubled.

Correspondingly, in conjunction with industrial development, an estimate of as high as 15,000,000 tons/year of cargo throughput is considered possible. Transport of this cargo, which mainly consists of export oriented products, will bring a tremendous increase in maritime transport demand and the Port of Mersin will be reinforced as Turkey's export gateway to the world.

With the developments in logistics functions, the position of Mersin will become more able to establish a land bridge between the Mediterranean, the Middle East countries, North Africa and the most important of the Euroasian countries around the Caspian and the Black Sea regions. With the possible introduction of the Baku-Ceyhan pipeline to transfer Caspian Oil to the Mediterranean the region will receive greater status and the Port of Mersin will be at the centre of maritime transport activities. In the context of these developments, it is more likely that the port will become a hub port in the near future and further studies may concentrate on organisational strategies in this area.

Being Turkey's largest port, backed by Turkey's greatest free zone; Mersin is a candidate for a leading port not only in Turkey's foreign trade but also in the international trade activities of the whole Eastern Mediterranean and Euroasia.

References

Bridger, G.A. and Winpenny, J.T. (1996) *Planning Development Projects, A Practical Guide to the Choice and Appraisal of Public Sector Investments*, Overseas Development Administration. HMSO: London.

Cargo News (1999) February.

Chadwin, M.L., Pope, J.A. and Talley, W.K. (1990) *Ocean Container Transportation*, Taylor and Francis: New York.

Containerisation International Year Book (1998) Black Bear Press Ltd: London.

Drewry (2000) *Mediterranean Container Ports and Shipping: Traffic Growth versus Terminal Expansion - An Impossible Balancing Act?* Drewry Publications: London.

Dynamar-Dyna Liner (1997) Weekly Magazine.

Hoffmann, J. (1998) *Concentration in Liner Shipping: Its Causes and Impacts for Ports and Shipping Services in Developing Regions*.

IAPH - International Association of Ports and Harbors (1999) *Biennial Report on Ship Trends-1999*, Committee on Ship Trends: Tokyo (Unpublished).

International Transportation Handbook '96 (1997).

Mersin Chamber of Shipping (1998) *Port of Mersin*, Mersin Chamber of Shipping Publication.

Mersin Chamber of Shipping (2000) Interviews with executives.

Ministry of Transport Japan, International Cooperation Agency (1996) *Containerisation and Liner Shipping Global Alliance and Port Management*, Seminar on Port Administration and Management, Tokyo.

OCDI-The Overseas Coastal Area Development Institute of Japan (1996) Container Terminal Planning, *Port Planning and Development*, Tokyo. (Unpublished).

Ocean Shipping Consultants Ltd. (1995) *The World Container Port Market to 2010*, Ocean Shipping Consultants Ltd: London.

Okada, H. (1988) *Port Development in Japan*, Overseas Coastal Area Port Development Institute of Japan. Tokyo (Unpublished).

TCS-Turkish Chamber of Shipping (1999) *Deniz Sektoru Raporu 1998*, Chamber of Shipping Publication: Istanbul.

Tomanbay, M. (1998) *Turkey's Water Potential and Southeast Anatolia Project*, Ottawa: Carlton University, Lecture Notes, October 29-30.

Toyran, R. (1997) *A Port Feasibility Study which can be Integrated with GAP (Southeast Anatolia Project)*, Istanbul University, Department of Maritime Transport. Unpublished MSc Thesis.

http://abone.turk.net/abide/med.htm

http://www.foreigntrade.gov.tr/SB/freezone.htm

http://www.dsi.gov.tr/gap.htm

http://www.eclac.cl/english/research/dcitf/lcg2027/contents.htm

5 Environmental Protection of Straits and the Law of the Sea with Particular Reference to the Mediterranean

D. BAŞER (KIZILSÜMER)
FACULTY OF BUSINESS,
DOKUZ EYLÜL UNIVERSITY, IZMIR
S. Ö. BAŞER
SCHOOL OF MARITIME BUSINESS AND MANAGEMENT,
DOKUZ EYLÜL UNIVERSITY, IZMIR

Introduction

The main purpose of this paper is to examine the provisions for environmental protection of straits in the law of the sea, an issue with particular relevance to Turkey and the internationally reputed Bosphorus.

For centuries, the subject of straits has been strictly connected to political and military concerns. However, straits being narrow waterways, sometimes with a congested traffic flow, under continuous threat from environmental disasters, have lately become a serious subject of marine environmental law. The risk of casualties such as collisions, strandings and groundings, is higher in straits than many other parts of the sea. Pollution caused by an accident near the coastline involving oil tankers, nuclear powered ships or ships carrying noxious substances can be devastating.

In this paper, we examine the legal solutions we can find in the international law of the sea to help prevent such disasters happening. Do we have sufficient legal means to cope with such disasters that may happen in straits?

Having examined the *travaux préparatoires,* during the Law of the Sea (LOS) Conference III the main aim of nation states was to reconcile the conflicting interests of strait states and user states. Whereas user states' interests lie in the freedom of passage in the high seas, strait states were looking for greater protection of the environment.

The paper comprises three sections. In the first we concentrate upon the evolution of the straits regime in the law of the sea and examine the case law (Corfu Channel case, 1949) and treaties (the Convention of Territorial Sea and Continuous Zone, 1958; the Law of the Sea Convention, 1982) that provide for a straits regime. Although the principle is the freedom of passage through straits, the regime governing straits has been in evolution since the Corfu Channel case.

Secondly, we deal with the provisions of the Law of the Sea Convention 1982 (1982 LOS Convention) covering the environmental protection of straits. The 1982 LOS Convention sets some duties for ships passing through straits and grants some regulatory and enforcement powers to the strait states. The 1982 LOS Convention generally refers to other Conventions prepared by the IMO on the safety at sea and prevention of marine pollution from ships.

Maritime traffic regulation, with particular reference to traffic separation schemes, can have a crucial effect on reducing the number of accidents in straits and this is also examined in this section.

The enforcement powers of strait states are regulated in Article 233 of the 1982 LOS Convention which provides for an exceptional right to take appropriate measures to prevent pollution.

Whilst law of the sea conventions have many articles on the protection of the environment, which is a principle of international law, straits are not protected in the same way as other parts of the sea with regard to regulatory and enforcement powers of strait states.

The Evolution of a Straits Regime with Regards to Environmental Protection

'Strait' is not defined in any of the Conventions produced by the UN Conferences on the Law of the Sea. Its ordinary meaning describes a narrow, natural passage or arm of water connecting two larger bodies of water. It is the legal status of the waters constituting the strait and their use by international shipping, rather than any definition of 'strait' as such, that determines the rights of strait and user states (Churchill and Lowe, 1988).

The Oxford Dictionary defines a strait as a 'comparatively narrow water-way or passage connecting two large bodies of water' and points out that, when used as a proper name, the word 'strait' is usually plural though with a singular sense (Bernhardt, 1989).

Being narrow waterways and thus choke points for maritime traffic, straits often require special environmental protection. There are also commonly conflicting interests between user and strait states. Whilst user states' interests lie in the free flow of transportation, the strait states have mainly environmental concerns. In spite of the importance of straits for international navigation and although several treaties had been concluded for navigation of a number of important straits, there was no clear concept of an 'international strait' until the Corfu Channel case was decided in 1949 (Anand, 1986; Brown, 1994; Koh, 1987; Mangone, 1987).

Corfu Channel Case[1]

The Corfu Channel is a strait, bounded on the west by the island of Corfu, belonging to Greece, and on the east by the mainland of Albania. In 1946, British Naval Ships while passing through the strait, struck mines and as a result, 44 officers and men lost their lives and the two ships were seriously damaged. The British Government brought its complaints to the UN Security Council, and the Council recommended that the two Governments should immediately refer the dispute to the International Court of Justice (ICJ).

The key passage in the Court's Judgement is as follows:

"It is, in the opinion of the Court, generally recognised and in accordance with international custom, that states in time of peace, have the right to send their warships through straits used for international navigation between two parts of the high seas without the previous authorisation of a coastal state, provided that the passage is innocent. Unless otherwise prescribed in an international convention, there is no right for a coastal state to prohibit such passage through straits in time of peace."

The case established that as a matter of customary law, warships had the right of innocent passage through international straits which could not be prohibited by the strait state. In addition, strait states may not make subject the exercise of the right of innocent passage to a requirement of previous authorisation from the strait state. This judgement of the case is worth mentioning because it explicitly points out that there is no requirement of previous authorisation for warships.

---

[1] Corfu Channel Case (UK, v. Alb.), 1949, ICJ 4 (Apr.).

Previous authorisation as well as notification could enable the strait state to take some preventive measures in case of dangerous passage that could cause an environmental incident.

Churchill and Lowe (1958) stated that:

"in the Corfu Channel case the ICJ judged that each state was under an obligation 'not to allow knowingly its territory to be used for acts contrary to the rights of other states'."

It could be argued that, taking the principle in the Corfu Channel Case:

"there is a general rule of customary international law that states must not permit their nationals to discharge into the sea that which could cause harm to the nationals of other states" (Churchill and Lowe, 1988).

1958 Geneva Convention on the Territorial Sea and Contiguous Zone (1958 Convention)

Article 16(4) of the 1958 Convention reads as follows:

"There shall be no suspension of innocent passage of foreign ships through straits which are used for international navigation between one part of high seas or the territorial sea of a foreign state."

From the wording of Article 16(4) it could be said that international straits are subject to the regime of innocent passage with the exception that innocent passage cannot be suspended.

The 1958 Convention was simply reflecting the decision in the Corfu Channel case; but extending the article to include not only straits connecting to areas of high seas but also straits connecting the high seas with the territorial sea of another State (Nandan and Anderson, 1989; Churchill and Lowe, 1988; Mangone, 1987). As the new formulation also refers to 'suspension' but not 'prohibition', it does restrict the rights of strait states compared with the Corfu Channel regime.

Although there was no provision specifically on the prevention of pollution in the straits, the article provides for 'non-suspendable innocent passage' which is more favourable to user states' interests than the innocent passage regime. As considered later in this paper, the non-suspendable innocent passage is still applicable in some international straits governed by Article 45(1)(b) and 38(1) of the 1982 LOS Convention.

The regime of non-suspendable innocent passage for straits is not controversial amongst commentators as it is a rule of international law However, these commentators could reflect an alternative interpretation for non-suspendable innocent passage. This alternative interpretation can be criticised as being very extreme. However, Article 16(4) should be read in connection with the previous paragraph (Article16/3) which states:

"the coastal state may...suspend temporarily in specified areas of its territorial sea the innocent passage of ships if such suspension is essential for the protection of its security."

So it can be concluded that non-suspension of innocent passage in the straits is only for security reasons. Strait states still have the right to suspend innocent passage for environmental concerns. In our opinion, non-suspension of innocent passage in the straits is only limited to security reasons because states could misuse it very easily claiming that every military vessel passing through the straits could threaten security and thus could endeavour to suspend innocent passage. This opinion could be supported from the *travaux préparatoires* of the 1958 Convention where the states negotiated on the requirement of previous authorisation or notification of warships.[2] The negotiation was only on the passage of warships as the strait states discussed this for security reasons and not for environmental concerns. In the Convention there is no provision providing for prior notification or authorisation either for warships or merchant ships.

Other provisions of the 1958 Convention on territorial seas are applicable to the straits that we will consider later. Otherwise, the single paragraph Article 16(4), is not enough for dealing with problems concerning straits - including environmental problems. Strait states have their rights provided in Article 16(1) which states:

"the coastal states may take necessary steps in its territorial sea to prevent passage which is not innocent."

---

[2] UN, UNCLOS I, OR, Vol..3, A/CONF.13/39.

Here, it must be proved that the passage is no longer innocent. Passage is innocent so long as it is not prejudicial to the peace, good order or security of the coastal state.[3] It is not stated in the article which activities are prejudicial to the peace, security and good order of the coastal state, however it is unequivocal that a damage or a threat of damage to the marine environment in the strait may render the passage non-innocent. The extensive wording of the Article 16(1) enables the strait states to regulate passage of ships through the straits and impose requirements for ships concerning environmental protection. The 1958 Convention does not provide for special articles for environmental protection of straits. However, Article 17 stipulates that:

> "ships shall comply with the laws and regulations enacted by the coastal state in conformity of these articles and other rules of international law, in particular with such laws and regulations relating to transport and navigation."

Other than the rules provided in the 1958 Convention, the strait states can make use of the customary law, the Corfu Channel case, the principle of environmental protection of the sea, state responsibility for damage to another state, and basic rules for legislative and enforcement powers of flag, coastal and port states for the protection of the environment in straits.

So here the provisions on protection of environment in straits is not different from provisions for the territorial sea and it is clear that environmental concerns noted during the 1970s are clearly reflected in the *travaux préparatoires* of 1982 LOS Convention.

1982 Convention on the Law of the Sea

The new Convention does not build upon the previous one; instead, it introduces an almost completely new regime. The core of the regime is the concept of transit passage (Scovazzi, 1995).

One of the most radical changes in the 1982 Convention was the codification of the transit passage regime. Generally, the regime of transit passage is more favourable to the interests of user states than was the previous regime of non-suspendable innocent passage. Transit may not be

---

[3] The 1958 Convention, Art.14(4).

suspended, nor can it be hampered by bordering states.[4] Moreover it applies to both ships and aircraft. The regimes which govern the straits were regulated in Articles 34-45 of the Convention.

Another innovation in the 1982 LOS Convention was adding Chapter XII for Protection and Preservation of the Marine Environment. The definition of straits is set forth in Article 37. Straits were defined as waterways:

"...used for international navigation between one part of the high seas or an exclusive economic zone and another part of the high seas or an exclusive economic zone."

Categories of Straits

The Conference attempted to make provision for navigation in all kinds of straits. The different categories of international straits provided in the 1982 LOS Convention, differ to some degree depending on the nature of the strait involved (Langdon, 1990; Schachte and Bernhardt, 1992/93).

Scovazzi (1995) pointed out that:

"it was done to preserve the application of the innocent passage regime to many straits."

Strait states generally are not satisfied with the provisions of the straits regime. They did not get enough protection, so they insisted on preserving innocent passage although it is non suspendable in some categories of straits.

Yturriaga (1991) admitted that:

"...this was done to overcome the resistance of states bordering various types of straits by excluding some of them from the application of the regime of transit passage."

The exceptions from the transit passage regime confirm that it is not widely supported. It did not meet the security and environmental concerns of the strait states and as a consequence, the regime of non-suspendable innocent passage has been retained for some kind of straits.

---

[4] 1982 LOS Convention Art. 44.

The categories of straits are as follows:

i. *Article 37 Strait*, the 'normal' international strait is, from a geographic vantage, the most frequently occurring strait of importance to international commerce and navigation. There are over 100 such international straits at present. The regime of transit passage applies in these straits. Under Article 37, this type of strait connects 'one part of the high seas or an exclusive economic zone and another part of the high seas or exclusive economic zone.'

ii. *Article 36* provides that if there is a strip of high seas or exclusive economic zone through a strait, and if navigation within that strip is as convenient as navigation through the territorial seas of the strait, then the ship or aircraft must avoid the territorial sea and remain in the high seas or exclusive economic zone portion. The regime of innocent passage will apply in those parts of the strait which lie within territorial sea limits and the regime of freedom of navigation through the exclusive economic zone or the high seas will apply in the middle section through the strait (Burke and De Leo, 1983).

Schachte and Bernhardt (1992/3) stated that:

"as a practical matter, this is meaningless at some point because the strait is no longer a strait, but high seas area in which freedom of navigation is applied, and the transit passage articles are inapplicable."

In these kinds of straits, environmental protection is subject to the high seas or innocent passage regime depending on the type of seas through which the ships are passing.

iii. *Article 38(1)* provides that:

"transit passage shall not apply if there exists seaward of the island, a route through the high seas or through an exclusive economic zone with respect to navigational and hydrographical characteristics."

The regime of non-suspendable innocent passage applies within the Article 38(1) type strait in the area between the mainland and the island. The most important example of this kind of straits is the Messina Straits, and as a consequence, it has become commonly named the Messina exception.

Non-suspendable innocent passage is much more favourable to the needs of the strait states because it sets out very precisely the concept of innocent passage by listing the activities in which ships should not engage during their passage through the territorial sea - in our case - the strait. As is stated in Article 19(2)(a):

"any act of willful and serious pollution contrary to this Convention renders the passage non-innocent."

As a result, ships passing through straits that violated this provision could no longer claim the right of innocent passage. In this type of strait, strait states can make use of all the provisions for innocent passage provided for environmental protection, except suspension of passage.

iv. *Article 45(1)(b) Straits*, addresses dead-end straits and also provides that the regime of non-suspendable innocent passage shall apply in an international strait:

"between a part of the high seas or exclusive economic zone and the territorial sea of a foreign state."

The points mentioned in the previous paragraph on the non-suspendable innocent passage also must be considered here.

v. *Article 35(c) Straits*, are regulated in whole or in part by long-standing international conventions that remain in force. These always apply to specific straits and examples include the Turkish Straits (Bosphorus), the Danish Belts and the Sound, and the Strait of Magellan.

The particular regime of international conventions in question applies to those straits. Article 35(c) adds nothing to these conventions, which are simply not affected by the 1982 LOS Convention, even though they provide for a different regime than transit passage. The 1982 LOS Convention does not make long-standing conventions *ipso-facto* applicable to those states which are not parties to them but are parties to the 1982 LOS Convention. These Conventions do nevertheless often accord the right of passage to third states, which consequently, are not entitled to exercise them (in the light of Article 36 of the Vienna Convention on the Law of the Treaties)[5] (Scovazzi, 1995).

What happens if these long-standing international conventions lack provisions on environmental matters? As the 'protection of the marine

---

[5] Vienna Convention Article 36 titled 'treaties providing for rights for third states'.

environment has become a basic principle of international law' (Spenskaya, 1986) the provisions provided in the 1982 LOS Convention can be used to fill the gaps in these treaties. To support this view, Jia stated that:

"such a regime (long-standing international conventions) includes both treaties in force and customary practice" (Jia, 1999).

The Right of Transit Passage

Transit passage is defined by Article 38(2) as follows:

"Transit passage means the exercise in accordance with this part of the freedom of navigation and over-flight solely for the purpose of continuous and expeditious transit of the strait between one part of the high seas or an exclusive economic zone and another part of the high seas or an exclusive economic zone."

It is a right akin to freedom of the high seas but for one purpose only - that of continuous and expeditious transit.

The new regime of transit passage is more favourable to the interests of user states than was the previous regime of non-suspendable innocent passage. Moreover, it applies to both ships and aircraft. Submarines, including missile-carrying nuclear submarines, are not required to transit on the surface.

Churchill and Lowe comment that:

"a general right of transit passage has not yet become established in customary international law. A right akin to transit passage does exist" (Churchill and Lowe, 1988).

Meanwhile, Jia stated that:

"Article 38(1) declared a right as if it is a matter of custom" (Jia, 1998).

Harris says that:

"The UK regards the right of transit passage as already stating customary law" (Harris, 1991).

Article 38(3) is silent about what constitutes an activity that is not an instance of transit passage. Transit passage accompanied by almost all of the non-innocent activities as enumerated in Article 19(2) cannot therefore be prevented, except when intrusions violating Article 39(1)(b) provoke self-defence in accordance with international law [6] (Jia, 1999).

Any activity which is not an exercise of the right of transit passage through a strait remains subject, under Article 38(3), to other applicable provisions of this Convention; Brown says that:

"these would include the provisions of Article 21, enabling the coastal state to adopt laws and regulations relating to innocent passage" (Brown, 1994).

However, this is debateable because regulatory powers of strait states are stated in Article 42 and strait states can only use their regulatory powers within the remit of Article 42.

1982 LOS Convention and Environmental Protection of Straits

One of the innovations in the 1982 LOS Convention was to include a new Chapter on Environmental Protection. In addition, throughout the Convention there are some provisions on environmental protection.

Having examined the *travaux préparatoires* of the 1982 LOS Convention it is clear that states seek to strike the right balance between the interests of the international communities on the freedom of navigation and the environmental and other interests of the strait states. However, it is important to consider whether they managed their objective of reaching a fair balance of conflicting interests or not. In this part of the study, we shall scrutinise the provisions of the 1982 LOS Convention dealing with environmental protection of straits. Article 39(2)a,b; Article 42(1)a,b,c and Article 43 directly regulate the environmental protection of straits.

[6]Jia, p.150.

Regulatory Powers of Strait States

Article 42(1) of the Convention states that, subject to the provisions of the section on transit passage, states bordering straits may adopt laws and regulations relating to transit passage through straits, in respect of all or any of the following:

- the safety of navigation and the regulation of maritime traffic, as provided in article 41;
- the prevention, reduction and control of pollution, by giving effect to applicable international regulations regarding the discharge of oil, oily waste and other noxious substances in the strait;
- with respect to fishing vessels, the prevention of fishing, including the stowage of fishing gears;
- the loading or unloading of any commodity, currency or person in contravention of the customs, fiscal, immigration or sanitary laws and regulations of states bordering straits.

Each of the four subparagraphs states the subject matter in which regulatory authority is granted to the strait states and gives a limiting factor to that authority (Yturriaga, 1991).

Malaysia, Morocco and Spain proposed major expansions of the regulatory competence of states bordering straits by adopting essentially verbatim, the list of competences of coastal states with respect to innocent passage in Part II, but none of these amendments were accepted by the Conference. The Convention, therefore, reduces the regulatory competence of the state bordering the Strait to the four above-mentioned subjects. This enumeration of the regulatory powers of the coastal state seems to have an exhaustive character.[7]

This restrictive and exhaustive enumeration of regulatory powers of strait states contrasts with the much broader catalogue included in the parallel Article 21(1) relating to innocent passage through territorial seas (Yturriaga, 1991). In contrast to the innocent passage regime, strait straits have no 'unilateral' regulatory power to impose sea lanes, TSSs, or other safety or pollution measures on ships in transit passage (Oxman, 1995). However, the states should be given the authority to enact legislation on

---

[7] ibid, p.172.

environmental matters. It is the strait states that can effectively evaluate the actual needs for protection of the marine environment.

Safety of Navigation and Regulation of Maritime Traffic

It is clear that if maritime traffic is sufficiently regulated, the possibility of accidents will be reduced.

Traditionally, shipping has been the most unregulated of all the transportation industries. However, it must be confessed that the unregulated state of marine transportation is reducing substantially. New systems of surveillance and control of navigation are being developed to improve marine safety. However, much remains to be done and marine safety has been brought to worldwide attention by major marine disasters involving collisions and groundings. The answer to marine accidents appears to lie in traffic control or 'routing' systems. Such control systems seems to be necessary particularly in areas of heavy traffic concentration, such as Dover, Gibraltar, Mallacca, Hormuz, and other straits (Gold, 1980a).

As we mentioned earlier, Article 42(1)(a) refers to Article 41 which provides for sea lanes and TSSs in Straits. From the wording of the Article 42(1)(a), Yturriaga (1991) pointed out that:

"the strait state's powers to regulate the safety of navigation and maritime traffic are limited to the designation of sea lanes and prescription of traffic separation schemes."[8]

Koh had the parallel view as he stated that:

"the power of strait states to adopt laws and regulations in respect of the safety of navigation and the regulation of maritime traffic is *severely* circumscribed. It can only be exercised as provided in Article 41 of the 1982 LOS Convention" (Koh, 1996).

Article 41 permits strait states to 'designate sea lanes and prescribe traffic separation schemes' where it is necessary to promote the safe passage of ships. While doing this, strait states must observe two conditions. The sea lanes and traffic separation schemes must conform to

---

[8] Yturriaga, p.285.

'generally accepted international regulations'. Here, the Article is referring to COLREG 72, and SOLAS. The sea lanes and traffic separation schemes must be submitted to and adopted by the relevant international organisation[9] which in this case is the IMO. The strait state cannot designate sea lanes or prescribe traffic separation schemes until they have been adopted by the IMO.

In our view, Article 42(1)(a) stating 'safety of navigation and regulation of maritime traffic', cannot be interpreted as safety of navigation is limited to sea lanes and the prescription of TSSs. In effect it is probable that regulation of maritime traffic refers only to sea lanes and traffic separation schemes while safety of navigation covers all the other vessel traffic services that will be scrutinised in a later section. The strait states must fulfil the requirement of adoption by the IMO only for the prescription of TSSs and the designation of sea lanes as provided for in Article 41.

The *travaux préparatoires* also support the view on the Draft article submitted by the United Kingdom,[10] Article 4(1)(a), and say that a strait state may make laws and regulations in conformity with the provisions of Article 3 which regulated the designation of sea lanes and prescription of TSSs. However, the 1982 LOS Convention, by explicitly stating 'safety of navigation and regulation of maritime traffic' tried to broaden the regulatory powers of the strait states.

Oxman (1995) can be quoted:

"It is important to emphasise that Article 41 itself, while referring explicitly only to sea lanes and TSSs, should be read broadly so as to permit full implementation under Article 41 of the full, substantive scope of the regulatory powers under Article 42(1)(a) that are to be implemented as provided in Article 41."

In other words the adoption of any measures consistent with the Convention deemed necessary by the strait state and IMO for safety of navigation and the regulation of maritime traffic. This broad reading of Article 41 advances coastal, maritime and environmental interests, any or all of which could be prejudiced.

It is hard to justify the view that placing all regulatory powers into the hands of the IMO would restrict the regulatory powers of the strait states.

[9] 1982 LOS Convention Art. 41(4).
[10] LOS Conference III, OR, vol.III, document A/CONF.62/C.2/L..3

Here, we must again mention that Article 41 merely gives the power to adopt sea lanes and TSSs by the IMO. The other laws and regulations on the safety of navigation can be decided by the strait state itself. Leaving the adoption of all safety measures to the IMO, severely restricts the regulatory powers of strait states. For example, a strait state that wants to establish a VTS should be able to enact its laws. It is merely the strait state that can decide on its needs and evaluate its capabilities. However, the strait state should be obliged to inform the IMO about the safety measures, or as occurs in territorial sea straits, the state could merely take into account the recommendation of the IMO. The 1982 LOS Convention lacks clear provisions about these measures for the safety of navigation. The safety navigation measures of which strait states are able to make use are indicated below.

Vessel Traffic Services (VTS)

Generally, Vessel Traffic Services is the term adopted by the IMO to describe the range of systems operated by coastal states over specific areas of sea adjacent to their ports or coasts under which ship traffic is subject to supply or exchange of information or the giving of advice or possibly, of instructions by coastal stations with a view to enhancing the safety and efficiency of that traffic[11] (Plant, 1990; de Bievre, 1990; Gold, 1983; Franckx, 1995).

The IMO Guidelines for Traffic Services, adopted by the IMO Assembly in November 1985 as a Resolution A.578(14), define Vessel Traffic Service as follows:

> "A Vessel Traffic Service is any service implemented by a competent authority, designated to improve safety and efficiency of traffic and the protection of the environment. It may range from the provision of simple information messages to extensive management within a port or waterway."

The reasons for establishing a VTS system may include the following:

---

[11] The notion of VTS is certainly not new. This concept was developed in Europe after the Second World War (Liverpool was the first port to introduce such a system in 1948) and later was introduced on the American Continent (the St. Lawrence River was the first area covered) and Japan (Tokyo harbour in 1972).

- assistance to navigation in appropriate areas;
- organisation of vessel movements to facilitate efficient traffic flow;
- the handling of data relating to ships participating in a VTS system;
- participation in action in case of accidents;
- and support of allied activities, such as pilotage and search and rescue.

These can be referred to as the VTS 'safety mandate'. Its broader aim is to prevent or minimise loss or damage to society as a whole by protecting coastal communities and the general public from the adverse affect of maritime accidents (Gold, 1983; de Bievre, 1990).

In June 1997 in an amendment to SOLAS, a new regulation on VTS was adopted. SOLAS Chapter V Regulation 8-2 sets out when VTS can be implemented.[12] It states VTS should be designed to contribute to the safety of life at sea, safety and efficiency of navigation and the protection of the marine environment, adjacent shore areas, worksites and offshore installations from possible adverse affects of maritime traffic. Governments may establish VTS when, in their opinion, the volume of traffic or the degree of risk justifies such services.[13]

Franckx pointed out that 'mandatory VTS appeared unacceptable in straits subject to the regime of transit passage' (Franckx, 1995). Reg. 8-2 SOLAS Ch.V, states that a VTS may only be mandatory in sea areas within a state's territorial water. However it is hard to justify this limitation for strait states. Although VTS is not explicitly mentioned in the 1982 LOS Convention, strait states should have the right of establishment and it is implied in the aim and scope of the Convention. The 1982 LOS Convention aims to protect the marine environment from disasters and VTS can contribute significantly to prevention of maritime accidents.

Gold (1983) provides a useful categorisation of VTS services. At present there are three common forms of marine traffic regulation in operation. These are: traffic separation schemes (TSSs), vessel traffic reporting systems (VTRS), and vessel traffic management systems (VTMS). The development of each of these modes predates the advent of

---

[12] web page, http://www.imo.org/assembly/857.htm
[13] webpage, www.imo.org

accidental marine pollution as a national and international problem. Environmental concerns have led a number of states, as well as the IMO, to apply or consider the application of these systems as a means of pollution prevention. How these systems are implemented and which type is applied to a particular area, will vary with the pollution problem presented. TSS is a passive method, while VTRS and VTMS involve interaction between coastal traffic centres and vessels in adjacent waters. Although there are significant differences between these passive and active modes of marine traffic regulation, many of the problems involved in their implementation and enforcement are similar. All three types contain departures from the tradition of the autonomy of a ship's master to sail a vessel on any route that he sees best.

The elements or functions of VTS can be stated as; vessel centre; coordinated information broadcasting services; traffic surveillance services; a TSS or other routing system; vessel clearance system; a ship movement reporting system (position fixing and track monitoring); navigational assistance services; pilotage services; services in support of allied activities e.g. search and rescue, pollution defence; traffic regulations (including legislation); and an enforcement system (Gold, 1983).

The most sophisticated and advanced system, Vessel Traffic Management System (VTMS), usually employs all the techniques of the other systems but supplemented with comprehensive radar surveillance and thus takes on many aspects of air traffic control (ATC). Surveillance provides information sufficient to monitor the presence and movement of all vessels in the control area at all times. The total information scenario, usually computer assisted, allows the coastal state's shore stations to regulate fully vessel traffic flow and would include, like air traffic control, advice on recommended courses and speed as well as information on hazards and dangers. Due to its high cost and multiple radar technology, VTMS at this time, is not suited to applications beyond ports and confined waters. Thus, VTMS presently is only found in ports and estuaries where traffic congestion and economic benefits of the smooth flow of traffic combine to justify such technology. Many European and North American, as well as some Japanese, ports and waterways use a variety of VTMS (Gold, 1982).

Traffic Separation Schemes

An important means of reducing the risk of collisions between ships is the use of TSSs to separate shipping in congested areas into one-way only lanes. The early examples of such schemes can be found in the voluntary agreements on routing made by shipowners trading on certain routes (for example in the China Sea) in the nineteenth century (Churchill and Lowe, 1988).

COLREG 72 Rules 1(d) and 10 define respectively, the competence of IMO to adopt TSS and the main technical regulations to be followed in this regard.

One of the most important innovations in COLREG 72 was the recognition given to TSSs. Rule 10, states that vessels using these schemes will be required to proceed in the appropriate traffic lane or zone. In so far as is practicable, vessels must avoid crossing traffic lanes. When crossing a lane is necessary, it must be accomplished as nearly as practicable at right angles to the general direction of the traffic flow.

COLREG 72 Rule 1(d) provides that TSSs may be adopted by the Organisation for the purpose of these Rules. Thus, it is not only Article 41 that gives competence to the IMO to adopt TSS, COLREG 72 provides for the same competence as well.

In addition, SOLAS Convention Ch.V, Reg. 8(b) recognises the IMO as the only international body competent to prescribe TSSs.[14]

Jia pointed out that:

> "It seems proper to observe that by Article 41, the LOS Convention has reflected existing practice subject, of course, to the right of transit passage"[15] (Jia, 1998).

There are a number of examples of TSSs and routing systems. The Sub-Committee approved proposals put forward by Indonesia, Malaysia and Singapore to extend the existing TSSs in their straits, to introduce three new TSSs, to establish two additional deep water routes, and to

---

[14]In SOLAS Convention Dh. V, Reg.(b) says that: 'The organisation is recognised as the only body for developing guidelines, criteria and regulations on an international level for ship's routing systems. Contracting Governments shall refer proposals for the adoption of ship's routing systems to the Organisation.'

[15] The important res. A.284 (VIII) Nov.1973.IMCO, Assembly, 8th sess., 128-137, setting out general rules for the establishment of TSSs, was adopted by IMCO's 84 member states. The rules are very close to those in Art. 41.

establish three Inshore Traffic Zones (ITZ). Meanwhile, the Sub-Committee for Safety of Navigation approved the establishment of a mandatory ship reporting system in the Straits of Malacca and Singapore.[16]

Article 22(3)(a) says that:

"In designation of sea lanes and the prescription of TSSs in territorial seas, the coastal state shall take into account the recommendation of the competent international organisation."

Here, the IMO need not give its consent to the scheme, approve or adopt it. This is different from the straits' provision that the IMO should give its consent on (adopt) the TSSs.

Another difference is that the coastal state may require tankers, nuclear powered ships and ships carrying nuclear or inherently dangerous or noxious substances or materials to confine their passage to such lanes.[17] There is no parallel requirement for tankers, nuclear powered ships and ships carrying nuclear or inherently dangerous or noxious substances or materials in transit passage, only a general obligation that ships in transit passage shall respect applicable sea lanes and TSSs established accordance with Article 41.

Prevention of Pollution

The regulatory powers of strait states on prevention of pollution are regulated in Article 42(1)(b) of the LOS Convention. It says that:

"Strait states may adopt laws and regulations relating to the prevention of pollution by giving effect to applicable international regulations regarding the discharge of oil, oily wastes and other noxious substances in the strait."

The main restriction contained in paragraph 1(b) of Article 42 is that it limits the regulatory powers of the strait state to give effect to 'applicable international regulations', and regarding the discharge of 'oil, oily wastes and other noxious substances'.

What does this provision imply while referring to 'applicable

---

[16] Safety of Navigation 43rd Sess., 14-18 July, 1997, web page, http://www.imo.org/meetings/ nav/43/nav/4 31.htm.

[17] 1982 LOS Convention Art. 22(2).

international regulations'? Jia (1998) attempted to answer this query, viz:

"Applicable international regulations refer to rules and standards which are binding only for a state which have bound by such regulations contained in an international treaty. The treaty must be in force. The use of this restrictive term means that a regulation concerning preservation of the marine environment, however reasonable and widely supported by the international community, cannot be applied to a state and to the ships flying its flag if the international treaty in which it is contained has not entered into force or even if the state concerned is not a party to the treaty."

From the wording 'applicable international regulations', binding treaties and conventions must be understood. However, treaties or conventions must be binding only for strait states adopting laws for prevention of pollution and not for user states. After being a party to a treaty, a strait state has a right to legislate its requirements into its own domestic laws.

However, it is useless to put such a restriction in the article. In some countries like Turkey, the Constitution provides that:

"If the Turkish Republic is a party to a treaty, the provisions of this treaty will be in force like laws."

As a result, Article 42(1)(a) provides nothing more than a Constitutional rule for Turkish law and this provision is nothing more than the repetition of a general rule that states can absorb the requirements of treaties into national law provided that appropriate constitutional procedures have been gone through.

If we interpret that this international regulation must also be 'applicable' to the user states, it will become very difficult to adopt laws on prevention of pollution.

Oxman (1995) believes that:

"The discharge standards that are 'applicable' to foreign ships in transit passage under the Convention are those identified in the chapter itself, namely in Article 39(2)(b), which requires ships in transit passage to comply with generally accepted international regulations, procedures and practices for the prevention, reduction and control of pollution from ships."

We would prefer to interpret the Article in the way that Oxman does so that its scope of application can be widened; however, with the way that the Article is worded, it is difficult to do this.

The regulations in question regard only 'the discharge of oil, oil wastes and other noxious substances in the strait.' Accordingly, strait states may not adopt regulations concerning the prevention, reduction and control of pollution resulting from causes other than discharges, which excludes pollution caused by dumping or resulting from a maritime casualty (Yturriaga, 1991). Spain proposed a series of amendments to the negotiating texts and to the Draft Convention itself; namely, to change 'applicable' to 'generally accepted' international regulations and delete 'oily' before wastes.[18] As a result of the pressure exerted by the president of the Conference, Spain agreed not to delete the word 'oily' before 'wastes', but insisted in maintaining the rest of her amendment to Article 42 (Yturriaga, 1991). The purpose of deleting the word 'oily' was to make the text broader by including all kinds of wastes.[19] This provision restricts the regulatory powers of strait states which are left without adequate powers to prevent pollution.

Other Regulatory Powers of Strait States

The strait state has a right to prohibit fishing in the straits from vessels in transit passage. If such a prohibition is violated, the activity of the fishing vessel may be classified as not being an exercise of transit passage with all the resulting consequences. The strait state can enact laws on prohibition of fishing and enforce them. The loading or unloading of any goods, currency, or persons in violation of customs, fiscal immigration or sanitary laws and regulations must be regarded the same as fishing. However, while enacting a law or enforcing it, the strait state must act in good faith and not abuse its rights stemming from the 1982 LOS Convention.

---

[18] UNCLOS III, OR, vol. XVI 169, meeting, paras, 1-4, and 6.
[19] LOS Conference III OR, vol XVI, Plenary meetings, paras.39 and 41.

Duties of Ships in Transit Passage

Article 39(2) says that ships in transit passage shall:

> "Comply with generally accepted international regulations, procedures and practices for safety at sea, including the international Regulations for Preventing Collisions at Sea; comply with generally accepted international regulations, procedures and practices for the prevention, reduction and control of pollution from ships."

In this Article, the 1982 LOS Convention brings about obligations for ships in transit passage. These obligations concern safety at sea and marine pollution protection from ships. These obligations can be in the form of regulation, procedure and/or practice. While regulations and procedures are written law, practices are customary law. However, this formulation of the Article is not very clear and leaves room for interpretation.

Scholars have severely criticised this provision. Brown (1994) for example, questions:

> "How one is to determine when a Convention aiming at universality becomes generally accepted, is not clear."

Jia trying to clarify the meaning of 'generally accepted' international regulations, refers to rules and standards recognised by international customs or embodied in international treaties, irrespective as to whether they are in force or not, provided that they have received general acceptance by the international community (Jia, 1998). They might be standards accepted by other states generally or at least by the majority maritime states which are most significantly affected by them and established in a Convention but not ratified by the particular states in question (Yturriaga, 1991). Although a strait state is not a party to the conventions dealing with safety at sea or pollution prevention from ships, the rules incorporated into this convention will be applied to the ships passing through the straits, and whatever the flag state's position to the Convention, these generally accepted Conventions will also be applied to ships in transit passage. Churchill and Lowe were in favour of the provision as they pointed out that:

"The duty to comply with international safety and pollution standards is independent of coastal legislation. The advantage of implementing these standards in such legislation is that they become directly enforceable by the coastal state authorities" (Churchill and Lowe, 1988).

It is hard to justify this provision because sometimes the straits require special environmental protection and what will happen then? One opinion is that strait states should be given all the powers to regulate the passage with few limitations such as has been done in the innocent passage regime.

The 1982 LOS Convention, in its articles on protection of the environment, directly refers to the international conventions on safety at sea and pollution from ships.

There is no corresponding reference in the section dealing with innocent passage, though it is provided that 'any act of wilful and serious pollution contrary to this Convention' will render the passage non-innocent, and the ship is required to comply with laws made by the coastal state under 21(1)(f) 'for the preservation of the environment of the coastal state and the prevention, reduction and control of pollution.'

The violation of Article 39(2)(b), by breaching the IMO Conventions, will deprive a ship of the right of passage. However, Articles 233 and 42(5) may be applicable (Churchill and Lowe, 1988).

International Conventions for Safety at Sea

The 1982 LOS Convention does not name the organisations which can create international regulations. However, the competent international organisation on these matters is generally regarded to be the IMO.[20] The 1982 LOS Convention also intends that the IMO's work will be supplemented by regional effort on matters appropriate for regional action, whether through the regional seas programme of the UN Environment Programme or other regional organisations (Oxman, 1995; Caminos, 1987).

---

[20] The IMO was established in 1958, when its constitutive treaty signed in 1948, came into force: until 1982 in was known as IMCO - the Intergovernmental Maritime Consultative Organisation. It has wide competence in matters affecting shipping and has adopted a detailed and technical approach to its work. Its committees such as the Maritime Safety Committee, the Legal Committee, and Maritime Protection Committee, have played a predominant role in drawing up regulations concerning navigation and pollution.

Ships exercising their right of transit passage through straits must observe the Regulations, regardless of whether the flag state or the strait state is a party to the International Regulations for Preventing Collision at Sea (COLREG), 1972[21] as well as other Conventions on safety;[22] SOLAS 1974;[23] CSC, 1972;[24] INMARSAT, 1976;[25] STCW, 1978.[26]

However, these Conventions rarely provide for special environmental measures for straits. They deal with safety of navigation in general terms; however straits being narrow waterways and sometimes surrounded by populated cities vulnerable to accidents, require special environmental protection.

COLREG 1972

This Convention groups provisions into sections dealing with steering and sailing; lights and shapes and sound and light signals. There are also four Annexes containing technical requirements concerning lights and shapes and their positioning; sound signalling appliances; additional signals for fishing vessels when operating in close proximity, and international distress signals.

Guidance is provided in determining safe speed, the risk of collision and conduct of vessels operating in or near TSSs. Other rules concern the operation of vessels in narrow channels, the conduct of vessels in restricted visibility, vessels restricted in their ability to manoeuvre, and provisions concerning vessels constrained by their draught.

In the 1981 amendment, a number of rules are affected but perhaps the most important change concerns Rule 10, which has been amended to enable vessels carrying out various safety operation, such as dredging or surveying, to perform these function in traffic separation schemes.

---

[21]Convention on the International Regulations for Preventing Collision at Sea, entry into force, 15.7.1997; num. of cont. parties, 132; world tonnage, 96.67%.

[22] Summary of status of Conventions as at 1 May 1999 updated from website, www.imo.org

[23] International Convention for the Safety of Life at Sea, entry into force, 25.5.1980; num. of cont. parties, 138; world tonnage, 98.45%;

[24] International Convention for Safe Containers, entry into force, 6.9.1976; num. of cont. parties, 65; world tonnage, 61.25%;

[25] Convention on the International Maritime Satellite Organisation, entry into force, 16.7.1979; num. of cont. parties, 86; world tonnage, 92.98%;

[26] International Convention on Standards of Training, Certification and Watchkeeping for Seafarers, entry into force, 28.4.1984; num. of cont. parties, 133; world tonnage, 98.11%.

The 1989 amendments concern Rule 10 and are designed to stop unnecessary use of the inshore traffic zone.[27]

The SOLAS 1974 Convention

The main objective of the SOLAS Convention is to specify minimum standards for the construction, equipment and operation of ships, compatible with their safety. Flag states are responsible for ensuring that ships under their flag comply with its requirements.

Control provisions also allow contracting Governments to inspect ships of other contracting states if there are clear grounds for believing that the ship and its equipment do not substantially comply with the requirements of the Convention. However, the reference in Article 39(2)(a) and (b) to generally accepted international regulations, does not enable strait states to use such enforcement powers as it only requires ships passing through straits to comply with navigational standards regulated by generally accepted international regulations.

Chapter V of the Convention deals with safety of navigation.[28] As provided in SOLAS Ch.V Reg.8-1, states can introduce mandatory ship reporting systems in order to improve safety at sea, the safety and efficiency of navigation and/or increase the protection of the environment. Once a system is adopted by the IMO, ships entering areas covered by ship reporting systems are required to report to the coastal authorities, giving details of their position and identity, as well as such supplementary information that has been justified in the proposal for adoption of the system. This may include for example, the intended movement of the ship through the area covered by the system, any operational defects or difficulties affecting the ship and the general categories of any hazardous cargoes on board (Morika, 1987). Such provisions of the SOLAS Convention place some burden on the strait states rather than flag states and to establish such a ship reporting system can be very costly and difficult to maintain.

As we have already seen, the Convention generally brings about obligations to the Contracting parties, so that first of all, strait states must be a party to the SOLAS Convention to apply its provisions. Strait states

---

[27] Focus on IMO, a Summary of IMO Conventions, February 1998, updated from, website,www.imo.org.

[28] SOLAS Convention Ch.V, reg.14.

are obliged to fulfill the requirements of the 'generally accepted international regulations' to benefit from their provisions providing for protection of the marine environment. If the strait state cannot provide facilities for surveillance of straits, then they cannot make use of some provisions of the treaties.

A number of amendments to the Convention exist; the May 1995 amendments affect Regulation 8 of Chapter V. The Regulation was amended to make ships' routing systems compulsory. Governments are responsible for submitting proposals for ships' routing systems to the IMO in accordance with amendments to the General Provisions on Ships' Routing which were adopted at the same time.

Regulation 8 of Chapter V is in conformity with Article 41 however, but goes on to emphasise that the ships' routing system is compulsory. It is thus much more favourable to the needs of strait states.

Conventions on Marine Pollution from Ships

MARPOL 73/78 is a combination of two treaties adopted in 1973 and 1978 respectively. It covers all the technical aspects of pollution from ships, except the disposal of waste into the sea by dumping, and applies to ships of all types.

A new and important feature of the 1973 Convention is the concept of 'special areas' which are considered to be so vulnerable to pollution by oil that oil discharges within them have been completely prohibited with minor and well defined exceptions. The main special areas are the Mediterranean Sea, the Black Sea, the Red Sea, and the Gulf area. Such straits must be protected under a special regime and so the respective states must make use of the 'concept of special areas' to protect these locations.

These provisions can be applied by strait states to ships passing through straits whatever their flag states' position with respect to the treaties. Strait states can apply them even if they are not a party to the Convention.

Enforcement Powers of Strait States and Liability

The only provision on enforcement powers of strait states which appears in Part XII relating to the Protection and Preservation of the Marine

Environment is in Article 233. It has the title of 'safeguards with respect to straits used for international navigation.'

Article 233 provides that nothing in sections 5, 6, and 7 affects the legal regime of straits used for international navigation. However, Article 233 brings an exception to the general rule of Article 233:

> "If a foreign ship other than which has sovereign immunity, violates the laws and regulations referred to Article 42(1)(a) and (b) causing or threatening major damage to the environment of the straits, the states bordering the straits may take appropriate enforcement measures."

This provision confirms the general rule that strait states are not granted enforcement powers within straits under any circumstances, or in relation to any matter over which they may have regulatory power, except where a violation of Article 42(1)(a) and (b) occurs. The specific cross-reference to Article 42(1) (a) and (b) limits the applicability of enforcement safeguards in Section 7 of Part XII only to violation of laws and regulations on safety of navigation or pollution control (Yturriaga, 1991).

However, Section 5 provides for international rules and national legislation to prevent, reduce and control pollution of the environment, Section 6 provides enforcement and section 7 provides safeguards. The designation of the Article is very complicated and even the *travaux préparatoires* does not give adequate indication of the meaning of Article 233. Only the statement relating to Article 233 of the Draft convention on the law of the sea in its application to the straits of Mallacca and Singapore[29] reflects the common understanding regarding the purpose and the meaning of Article 233 and its application to those straits. The Article is widely considered as not properly drafted, and it lacks precise provisions on the enforcement powers of strait states.

So what happens if a foreign ship violates Article 39(2) (a) and (b)? In this Article, it states that:

> "Ships shall comply with generally accepted international regulations on safety at sea and prevention from pollution."

So how shall a strait state respond to violations other than provided in

---

[29] UNCLOS III, OR, vol.XIV, 182ND meeting documents.

Article 42(1)(a) and (b)? It is hard to justify the rationale of this limitation.

At this point we will scrutinise firstly, the circumstances under which a foreign ship that violated the laws and regulations under Article 233 would be deemed to be causing or threatening major damage to the marine environment and secondly, what can be the appropriate measures to be taken and do they include prohibition, suspension or hampering of passage?

Koh considers that 'threaten' must involve a threat of major damage to the environment. There will always be grey areas whether the threatened damage is major or minor. The evaluation is much more difficult if no damage has yet occurred. It would be a speculative affair. To this extent the provision is very open-textured (Koh, 1980). Scovazzi said that 'the wording of Article 233 is inevitably vague. Expressions such as 'major damage' are open to different interpretations' (Scovazzi, 1995).

Caminos is of the view that:

"The strait state may subjectively determine the probability of the 'threat' to happen as well as the amount of damage foreseeable to the marine environment. The criteria 'threat' and 'major' are theoretical and open to different interpretation" (Caminos, 1987).

Does, for example, a VLCC that has an under keel clearance of say, three metres or which cruises at 13 knots, threaten major damage to the marine environment? The 'IMCO Rules' lay down minimum standards for safe navigation, then by parity of reasoning any violation, however minor, would be a threat of some kind (Koh, 1980). If it is an oil laden vessel of over 100,000 dwt, an oil spillage could cause major damage. Of course, if it is empty or if it is carrying cargo which when emptied into the sea would not cause damage to the marine environment, any infringement of the IMCO Rules will not come within the ambit of the section (Churchill and Lowe, 1988).

As mentioned above in the scholars' views, to decide on what is a 'threat' and 'major', requires discretion. Divergent interpretation between strait states and user states can cause disputes.

The wording of the Article necessarily restricts its applicability to major damages. In our view, in this article 'major' unnecessarily restricts the enforcement powers of strait states. Maduro says that:

"When considering the fact that 70 or 80% of oil emanating from tankers is the result of routine discharges in normal tanker operations, which from a single vessel can hardly be considered as a major threat, ...for this reason, this provision is clearly inadequate for dealing with this primary source of pollution" (Maduro, 1980).

It is hard to justify the limitation of 'major' threat or damage with the aim of trying to protect the environment. It restricts the enforcement powers of strait states as they have only enforcement powers in the exceptional cases of major threat or damage.

The second problem to be dealt with is the scope of the enforcement measures that can be taken by strait states. The appropriateness of a state's enforcement jurisdiction shall respect *mutadis mutandis* the provisions of section 7 of Part XII. Safeguards spell out measures to facilitate proceedings, physical investigations of vessels, monetary penalties and the like; basically those measures stipulated in Article 220 (Koh, 1980).

As Article 220 permits inspection of the vessel, its detention if necessary, and/or the institution of legal proceedings, one wonders whether this is the type of enforcement measures which Article 233 intends to apply to straits. The answer to this question in its context and in the light of its objective and purpose, would have to be affirmative (Koh, 1980).

Jia extended the applicable articles by stating that:

"The situs of Article 233 in the 1982 LOS Convention, the enforcement measures to therein can only be taken on the basis of Articles 216, 218, 220" (Jia, 1998).

As the wording of the article restricts the enforcement power of strait states to Section 7 and being Articles 216[30], 218[31] and 220[32] in Section 6 of Part XII., it is hard to agree with these same views.

It is only section 7, titled Safeguards, which could be applied *mutadis mutandis* to strait states. For that reason this article must be criticised for being too restrictive, and not giving the proper means to strait states to protect the marine environment.

The last problem to be dealt with is whether strait states could prohibit

---

[30] Article 216 titled 'enforcement with respect to pollution by dumping'.

[31] Article 218 titled 'enforcement by port states'. If we accepted that Art. 218 applies to straits, this strait should have a port in it to apply this provision.

[32] Article 220 titled 'enforcement by coastal states'.

or impair the passage of ships.

Scovazzi said that:

"Article 233 does not specify what enforcement measures are. At this point it is possible and preferable, to interpret this provision as allowing bordering states to forbid the passage of such ships" (Scovazzi, 1995).

Caminos is of the view that:

"Having regard to enforcement safeguards, states bordering straits should as far as possible, respect the general rule that transit passage cannot be impeded or impaired. When a violation of the safety of navigation standards or pollution regulations has occurred or is threatened, the strait state must act cautiously. It must evaluate the amount of damage or threatened damage that could happen. And it must judge on the reasonableness of the enforcement measures undertaken. These enforcement measures must be proportionate and not arbitrary" (Caminos, 1987).

Meanwhile, Koh is of the view that:

"A textual examination of Article 233 demonstrates that any enforcement acts taken under its authority are an exception to the general rule embodied in Article 42(2) that transit passage cannot be impeded. Any other interpretation would overwhelm the exception and render the pinpoint cross-reference to Article 42(1)(a) and (b), meaningless" (Koh, 1980).

So the scholars have the view that in the case of violation of Article 42(1)(a) and (b), transit passage can be impeded.

However, it seems that the scholars rather ignored the wording of the article as it says 'nothing in sections 5, 6, and 7 affects *the regime of straits...*' The regime of the straits should be understood as transit passage. The *travaux préparatoires* support the view that 'transit passage' must be understood from the regime of straits mentioned in Article 233[33] and the most important character of the transit passage is that it cannot be hampered. So it is difficult to match the views of the scholars with the

---

[33] III UNCLOS vol.XIV, OR, document, A/CONF/.62/WS/12.

wording of Article 233. With these ambiguities in Article 233 it is inevitable that there will be different opinions.

Article 25(1), entitled Rights of Protection of the Coastal State states:

"...the coastal state may take the necessary steps in its territorial sea to prevent passage which is not innocent."

No parallel provision exists in Part II of the LOS Convention relating to straits used for international navigation. It is difficult to justify why the 1982 LOS Convention lacks such a provision. Instead, although the coastal state enjoys some regulatory powers concerning transit passage, its law and regulations should not, in their application, have the practical effect of denying, hampering or impairing the right of passage.[34] This provision could be interpreted as that the strait state could never enforce any measures impairing or hampering passage. There should be an explicit provision that if it is required for environmental concerns the strait state has the right to impair the passage of ships through the straits.

Another restriction in Article 233 is that it does not permit enforcement measures to be taken with regard to ships and aircraft entitled to sovereign immunity, namely any ships, naval auxiliary, other vessels or aircraft owned by or operated by a state and used, for the time being, only on government non-commercial service.[35]

Even if such ships or aircraft are actually causing major damage to the marine environment, the role of bordering states is limited to waiting, taking note and presenting the bill to the flag state or registry state according to Article 42(5) (Scovazzi, 1995) which says that 'they shall bear international responsibility for any loss or damage which results to states bordering straits.' However, in the parallel provision for territorial seas in Article 30, the:

"...coastal state may require the warships to leave the territorial sea immediately if any warship does not comply with the laws and regulations of the coastal state and disregards its request for compliance."

In a statement by the Spanish delegation, they submitted that Article 233 has to be considered discriminatory against strait states, inasmuch as it

---

[34] 1982 LOS Convention Art. 42(2).
[35] 1982 LOS Convention Art. 236.

is precisely their geographical narrowness that creates greater risks of accidents which could cause irreparable damage to the marine environment.[36]

The provisions on these enforcement powers of strait states must be clear and precise otherwise these could cause many disputes between user and strait states.

In territorial seas, the coastal state is empowered to take inspection and 'cause proceedings, including arrest of vessel' where it has violated its laws and regulations relating to prevention, reduction and control of pollution. However, we cannot apply this provision because this provision lies in section 6 which as stated in Article 233, does not affect the legal regime of strait states (Maduro, 1980).

There is, however, a general duty to comply with all laws made within the legislative competence allowed under the Convention to Straits States.[37]

Liability

Under Article 42(4), foreign ships exercising the right of transit passage have to comply with the laws and regulations relating to transit passage adopted by a strait state. Failure to do so would render them liable to proceedings under such laws and regulations. It is just a general liability provision that does nothing more than provide a framework for liability problems.

Moreover there is no detailed provision provided for straits in Part III as provided in Part II for criminal jurisdiction of the strait state[38] or civil jurisdiction in relation to ships[39] in territorial seas.

It might also mean that the ship's passage would become an 'activity which is not an exercise of the right of transit passage'. It might be then subject to the regime of innocent passage, including the remedies available to the coastal state for breach of the conditions of innocent passage or of the laws and regulations of the coastal state relating to innocent passage (Brown, 1994).

---

[36] IIIrd UNCLOS vol.XIV, OR, document, A/CONF/.62/WS/12.
[37] Art.40.
[38] 1982 LOS Convention Art.27.
[39] 1982 LOS Convention Art.23.

Conclusions

Reconciling conflicting interests of strait states and user states has always been the aim of jurists. Earlier, it was the military and security interests, nowadays the economic and environmental interests of strait states have to be accommodated.

The Corfu Channel Case was not about environmental protection of straits, however, it entitled the user states to the general right of passage and detailed the usage of the right. There was an indication in the decision of the ICJ that dealt with the environmental protection of straits as it says 'each state was under an obligation not to allow knowingly its territory to be used for acts contrary to the rights of other states'.

The 1958 Convention did not provide for any provision on the protection of the environment in straits. As it says in Article 16(4), 'There shall be no suspension of innocent passage of foreign ships'. The regime regulated in this chapter, is much more favourable to the interests of user states when we compare it with the regime of innocent passage in territorial seas. However, it should be noted that provisions for innocent passage other than the right of suspension can be applied here. For example, foreign ships passing through the straits must comply with the laws and regulations enacted by the coastal state.[40]

The 1982 LOS Convention brought about a transit passage regime for certain categories as well as the new regime for environmental protection of straits. As it includes 'duties of ships and aircraft during transit passage'[41] and 'laws and regulations of states bordering straits relating to transit passage',[42] then the protection of the environment embedded into these articles are part of the transit passage regime. Non-suspendable innocent passage regime is retained for certain categories of straits, as well as the regime provided by the long-standing international conventions. However, what happens if the long-standing international convention lacks provisions on protection of the environment? Then, as we accepted the protection of the marine environment as a principle of law, the provisions granting protection rights to strait states of the 1982 LOS Convention must be applied to fill the gaps in those long-standing treaties.

The primary rules apportioning competence to the strait states are to be found in the 1982 LOS Convention while the secondary rules,

---

[40] Article 17, the 1958 Convention.
[41] Art.39
[42] Art.42

containing more technical rules and regulations, on the other hand are primarily to be found in the relevant conventions drawn up under the auspices of the IMO.

For the purpose of this study we took COLREG 1972, SOLAS 1974 and MARPOL 73/78 as generally accepted international regulations as well as the other international conventions. There are some provisions in these conventions for safety of navigation, regulation of maritime traffic and protection from pollution that can be useful while protecting straits from environmental disasters.

The regulatory powers of strait states are regulated in Article 42. However, these regulatory powers are restricted only to those mentioned in the article. The strait states should be given more regulatory powers. In this case, only a sentence like that provided for territorial seas can be enough as it states 'a strait state may adopt laws and regulations on the preservation of the environment and the prevention, reduction and control of pollution'.[43]

Regulation of Maritime Traffic has the effect of reducing the number of accidents in straits. Although Article 42(1) makes a cross reference to Article 41 which provides for sea lanes and TSSs, we came to the conclusion that safety of navigation and regulation of maritime traffic cannot be limited to sea lanes and TSSs. Providing for more sophisticated mechanisms, other VTS can more effectively deal with maritime incidents. We suggested that Article 42 is implying not only sea lanes and TSSs but also other VTS. However we reserved the provisions of other conventions providing for VTS. While prescribing VTS the strait states need not follow the procedure provided in Article 41 because VTS are not explicitly mentioned in Article 41. The 1982 LOS Convention lacks clear provisions on what the safety of navigation measures are. As a result, we gave information on other safety measures provided in the generally accepted international regulations.

The regime of transit passage is more favourable to the interests of user states than was the previous regime of non-suspendible innocent passage.

The enforcement powers of the strait states are regulated and restricted by Article 233. Part XII of the 1982 LOS Convention deals with 'Protection and Preservation of the Marine Environment'. However, Article 233 by saying that 'nothing in sections 5, 6 and 7 affects the legal regime of straits used for international navigation', excludes the strait

---

[43] Art.21(1)f.

states from enforcement powers that are the backbone of protection of the marine environment. This provision significantly restricted the enforcement powers of strait states. This provision is poorly drafted and provides inadequate enforcement powers for straits.

Primary References

The Law of the Sea, Straits Used for International Navigation.
Legislative History of Part III of the United Nations Convention on the Law of the Sea Volumes I and II. United Nations, New York, 1992.

References

Anand, R.P. (1986) Transit Passage and Overflight in International Straits, *International Journal of International Law*, 26.
Bernhardt, R. (1989) *Encyclopaedia of Public International Law*, 11, Law of the Sea, Air and Space, North Holland.
Brown, E.D. (1994) *The International Law of the Sea*, Volume I, Introductory Manual, Dartmouth.
Caminos, H. (1987) *Recueil des Cours, Collected Course of the Hague*, Dordrecht, Martinus Nijhoff.
Churchill, R. R. and Lowe, A. V. (1988) *The Law of the Sea*, Manchester University Press, 2nd ed.
De Bievre, A. (1990) Vessel Traffic Services and the Law, *Journal of Navigation*, 38.
De Yturriaga, J.A. (1991) *Straits Used for International Navigation, A Spanish Perspective*, Martinus Nijhoff.
Franckx, E. (1995) Coastal State Jurisdiction with Respect to Marine Pollution - Some Recent Developments and Future Challenges, *International Journal of Maritime Commerce and Law*, 10, 2, 253-280.
Gold, E. (1983) Vessel Traffic Regulation: Interference of Maritime Safety and Operational Freedom, *Journal of Maritime Law and Commerce*, 14, 1.
Harris, D.J. (1991) *Cases and Materials on International Law*, 4th ed., Sweet and Maxwell, London.

Jia, B.B. (1998) *The Regime of Straits in International Law*, Oxford, Clarendon Press.

Koh, K.L. (1980) *Straits in International Navigation-Contemporary Issues*, Oceana Publications Inc.

Koh, T.T.B. (1986) The territorial Sea, Contiguous Zone, Straits and Archipelagos under the 1982 Convention on the Law of the Sea, *Malaya Law Review*, 29.

Maduro, M.F. (1980) Passage through International Straits: the Prospects Emerging from the Third United Nations Conference on the Law of the Sea, *Journal of Maritime Law and Commerce*, 12, 1.

Mangone, G.J. (1987) Straits Used for International Navigation, *Ocean Development and International Law*, 18.

Nandan, S.N. and Anderson, D.H. (1989) Straits Used for International Navigation: a Commentary on Part II of the United Nations Convention on the Law of the Sea 1982, *British Yearbook of International Law*.

Neill, H.J. (1995) The Channel Navigation Information Service for the Dover Strait, *Journal of Navigation*, 43.

Oxman, H.B. (1995) Environmental Protection in Archipelagic Waters and International Straits - the Role of the International Maritime Organisation, *International Journal of Maritime Commerce and Law*, 10, 4.

Plant, G. (1990a) International Legal Aspects of Vessel Traffic Services, *Marine Policy*, 14, 1.

Plant, G. (1990b) International Traffic Separation Schemes in the New Law of the Sea, *Marine Policy*, 14, 5.

Plant, G. (1996) Navigational Regime in the Turkish Straits for Merchant Ships in Peacetime, *Marine Policy*, 20, 1.

Schachte, W.L. and Bernhardt, J.P.A. (1992/1993) International Straits and Navigational Freedoms, *Virginia Journal of International Law*, 33.

Scovazzi, T. (1995) Management Regimes and Responsibility for International Straits - with Special Reference to the Mediterranean Straits, *Marine Policy*, 19, 2.

Spenranskaya, L.V. (1986) Marine Environmental Protection and Freedom of Navigation in International Law, *Ocean Yearbook*, 6.

Sturt, R.H.B. (1984) *The Collision Regulations*, London, Lloyd's of London Press Ltd.

Treves, T. (1997) *The Law of the Sea, the EU and its Member States*, Martinus Nijhoff.

6 Crises in Ports and the Significance of Procuring Contingency Plans

HAKKI KIŞI
SCHOOL OF MARITIME BUSINESS AND MANAGEMENT,
DOKUZ EYLUL UNIVERSITY, IZMIR

Introduction

The aim of this study is to highlight the probable crises at ports in general and the ports in Turkey in particular that are regularly encountered and to put forward the means of minimising damages they are likely to bring about. The study will consider a crisis as the simplest definition, "a state of being complex" or "any unexpected occurrence breaking the established routine" with a further meaning of "a serious threat resulting in danger terminating port operation". The study will also consider the broad-scale effects of crises and the economic, social, environmental, regional and/or national losses that they bring. Considering the multi-faced feature and the diversified effects of crises in ports, the study proposes certain proactive measures and reactive means to achieve efficient restoration of services. The guidelines proposed herein will be based on the relevant national and international regulations and conventions as well as the actual situation of ports in Turkey with regard to the available approaches observed. The very first step of this study was to examine the relationship of national and international regulations and to develop a detailed questionnaire carried out within the ports in Turkey. The former aimed at establishing the overall frame of measures to be taken and the latter to draw a clear picture of the specific situation.

A crisis simply is any unusual, unexpected and unstable condition, which is likely to slow down, deform, diminish or even destroy any system in which it is involved. It might arise from external and/or internal factors. External disasters likely to affect ports comprise natural ones; sudden changes in the natural environment (Ross and Murdick, 1973);

earthquakes, e.g. Kobe on January 17th, 1995 (http://www.ege.com/2000); global economic ones, e.g. the Asian economic crisis as well as the one in Russia encountered quite recently; terrorism and wars. Internal inadequacies include shortage of up-to-date technology, incapability of top managers, lack of training and organisation might also comprise a suitable environment where crises can easily find roots. In addition, fires and strikes, which might stem from both internal and external factors, are likely to affect basic port operations. Crises differ greatly from many other common and potential problems in that they are extraordinary stresses which need to be handled immediately and resolved by several different approaches and can consist of stress, insecurity, uncertainty, anxiety and panic and also that they are both critical and threatening. As for the stages of crises, it is widely accepted that there are three: Prodromal/Build up, Acute/Breakout, and Abatement stages (Weitzel and Johnson, 1989). The first is the initial stage, the pre-alert which can provide the chance to reduce its impact; the second is the realisation stage, wherein stress and anxiety are maximised and relations with third parties, i.e. routine operations, tend to diminish (Summers 1977); and the third is the final stage of the crisis, the length and impact of which can vary depending upon the acts taken against the crisis.

The three stages of a crisis require three stages to minimise its harm. These are preparedness, response and restoration, which will be dealt with in detail in the next section. Meanwhile, the third section will focus on assessment of the present situation in the ports in Turkey in terms of proactive and/or reactive actions taken against any probable crises. In the last section, a detailed list of findings will be presented together with an evaluation of the actions available and used.

The objectives of this study, hence, are two-fold; generating an ideal contingency plan practicable in all ports, and the ones in Turkey in particular, and revealing an overall panorama of the ports in Turkey in terms of proactive and/or reactive actions taken against potential crises. When these two analyses are put together, the achievements and failings of Turkish ports in times of crisis should be apparent.

The methodology used in this study is based on three basic sources of gathering data. These are secondary data, an interview with the responsible staff at the Port of Izmir and a questionnaire study carried out with the ports in Turkey. The secondary data outlined the frame of a sample contingency plan for ports in general against any potential crises. The interview with the head of the Contingency Committee at the Port of Izmir both set an example for the general profile of the ports in Turkey concerning their approaches towards contingency planning and contributed to shaping the contents of the questionnaire. The questionnaire aimed to draw a picture of the actions taken against crises. A total of 16 questions were included. These questions aimed to reveal a clear picture of the contingency plans available in the ports. The questionnaires were faxed to 20 ports in Turkey, and 13 of them responded, which was enough for the researcher to obtain an overall picture.

Contingency Planning in Ports

Ports in general are technology and capital extensive foundations and in addition their operations are time-based. Any damage to their infrastructure is likely to result in not only slowing down and/or even diminishing operations but also extensive costs for restoration. Even a slight obstruction in any part of the whole chain of operations carried out at a port is likely to affect other sections of the chain. In fact, any failure with any part of this chain of operations tends to be decisive in interrupting transit. Such sensitivity to these problems requires efficient protection against potential dangers. For example, loading and/or discharging operations are based on some other prior operations such as pilotage, entry to the port, berthing and/or anchorage. Likewise, departure of a vessel is necessarily bound to certain prior operations such as handling services, storage, stowage, weighing, waste removal, fresh water supply, fuelling/bunkering, tugging and pilotage services, etc. This interconnection of services makes the reduction and mitigation of these problems that much more important.

Ports are vulnerable to many crises but the extent of their effects vary. The dangers likely to strike ports are fires, strikes, natural disasters like earthquakes and floods, wars, casual accidents, technological laxities or employee-sourced negligence. Along with these risks, political and national/international based regulation mishaps, global economic crises and even managerial inadequacies constitute another set of threats. Still another set of dangers stem from the lack of a well-organised contingency plan, or laxity in adequate compliance with an existing contingency plan.

Aspects of a Contingency Plan for Ports

A contingency plan for ports could be formed in three basic steps: preparedness, response, and restoration.

Preparedness (Pre-crisis Preparatory Phase)

This step comprises the actions to be taken prior to any crises. In other words, this step requires putting into effect a proactive brain-storming exercise and determines the actions to be taken accordingly.

This step is supposed to answer the following basic questions: (1) What might happen? (2) What damage might the crisis cause? (3) How accurately could the management simulate the crisis and the damage? (4) What facilities does the management have to make use of in fighting the likely crises? (5) What should be done to minimise the damage the anticipated crisis is likely to cause? (6) How could the scenarios be put into a detailed crisis management plan?

The Nature of the Contingency

What might happen? The answer to this question involves a detailed situation analysis, which also includes answers not only to what but also where, how and how much? In other words, the type, the place, the manner and the extent/power of the crises have to be anticipated and clarified.

Expected Damage

What damage might the assumed crisis cause? This question requires preparing a detailed risk analysis. Such an analysis is supposed to clarify/specify the sources of crisis and their costs. This analysis helps reveal the loss in capital, lives, and natural resources in the case of a particular crisis.

Scenario Plotting

How could the management simulate the crisis and the damage anticipated? The answer to this question involves generation of scenarios which are supposed to be as accurate and factual as possible.

Determination of the Facilities and Equipment for Fighting against Contingencies

What facilities are available to be put into service in combating the anticipated crisis? The answer to this question aims to identify these facilities and their logistics i.e. the contents of the inventories, the places they should be kept, how they should be handled and transported in case they are needed. A comprehensive and practicable data processing system is of vital importance. Meanwhile outsourcing any of the tasks likely to be involved should be taken into consideration.

Action Plans

What should be done to minimise the damage the anticipated crisis is likely to cause? The key answer to this question is well-organised task allocation. Who should do what, when, where and how needs to be specified. Such an allocation has to be arranged in such a way that the qualifications of the allocated staff should match the strict requirements of the task involved depending upon the feature of the crisis anticipated Again, comprehensibility and practicability are the key factors in such allocation.

Task Description and Designation of Responsible Staff

Designating a 'responsible leader' could help centralise and control the procurement of the tasks allocated. Although the present case at some ports seems to be that the top manager is designated as a responsible leader, this does not have to be so. The qualifications of the staff to be designated do have to match the relevant requirements in case of any emergency. In addition to designating a top responsible leader, each of the units involved in the port operations might designate a unit leader.

Staff Training and Scenario Testing

Training plays a crucial role in combating crises. On-the-Job training is to be accompanied by assessing the scenarios in practice. The more often and the more efficiently the tasks are exercised, the more efficaciously they will be put into effect when the relevant anticipated crisis hits the port. Such exercises are likely to reveal the failures, enhance the required success and contribute to testing the costs, parameters, applicability and laxness that may exist.

Generating Scenarios for a Crisis Management Plan

The overall aim of forming a crisis management plan is primarily to keep the crisis under control. Achieving this aim clearly requires effective coordination. Such coordination is to cover all the relevant links of communication, action plans, decisions and conformity with the national and international rules, regulations and conventions. Besides, "this plan should work under the foresight of strategic management." (Tüz, 1996) Furthermore:

> "the plan should include objectives, priorities for protection and saving, a sensitivity chart that depicts risk areas and levels....., categorising action levels, coordination and collaboration with other organisations, warning and monitoring systems and designing an organisation for intervention. An incident command system (ICS) should be developed where the incident command organisation has been described in terms of command, planning, logistics, operation and finance." (http://www.env.gov.bc.ca)

Response

This step is related to the acute phase of the anticipated crisis, the stage at which the institution is face to face with the actual crisis. It therefore, consists of the actions pre-specified in the crisis management plan. The actual practice of these actions/measures constitutes the backbone of this step.

This step of combating the crisis could be briefed as follows:

Perception: Perceiving the crisis, immediate recognition of its type, nature and probable power and getting ready to initiate the relevant plan based on this perception.

Apprehension: Reaching a closer recognition of the crisis faced and hence, getting more precise information about the plan to be initiated and the tasks to be activated.

Alarming and Communication: Based on the clear recognition and apprehension of the crisis faced, initiating/activating the relevant part of the crisis management plan, and setting into practice the predetermined communications links.

Coordination and Allocation: Activating the teams, tools, vehicles and the staff specified in the crisis management plan.

Logistics: Having access to the place(s) predetermined in the relevant part of the emergency plan and activating immediate transportation of the required tools and/or vehicles to the location of the crisis.

Search and Rescue: Securing for example, power supplies, first aid, extinguishing fires, etc.

Medical Aid: Reaching the wounded, activating medical aid facilities and/or activating the means of lessening damage and putting into effect all predetermined measures.

Damage Assessment: Making a rough and immediate assessment of the damage from the crisis.

Reporting: Reporting assessment of the damage to the relevant places/units prespecified.

Temporary Repair: Removing and making safe the damage.

Restoration

This step covers the things to be done after the crisis is over - related to the post-crisis phase - and includes cleaning, maintenance, replacement, reengineering, investment, and employment.

Since the focus of this study is on the first two stages of a crisis rather than the post crisis period, the factors that make up this phase will not be further clarified.

Crisis Management Plans in Turkish Ports

Ports in Turkey can be classified into three groups - governmental, municipal (operated by local governments) and private ports. The first group are operated by State Economic Enterprises. Some of these ports are operated by the Turkish State Railways and they are connected with the railway network. Examples include Haydarpaşa, Derince, Bandırma, Izmir, Mersin, Iskenderun and Samsun. Another group of governmental ports, such as Trabzon, Kuşadasi and Antalya, are operated by the Turkish Maritime Organisation.

These two State Economic Enterprises, Turkish State Railways and Turkish Maritime Organisation, operate independently, but are under the control of the Ministry of Transport. In the headquarters of both organisations, there is a separate Department of Ports responsible for planning and coordinating the ports under their authority. Each individual port is managed by a port manager appointed by the related State Enterprise.

All the ports operating under the command of these state enterprises are full-service ports and the services to ships and cargoes given by these ports are done so using their own employees and equipment. The second group of ports - municipal - is comparatively small and limited to a small volume of coastal traffic serving the local needs of provincial towns. The total number of such ports is 22.

The Questionnaire Applied in the Main Turkish Ports

A set of sixteen questions was directed to the ports. The questions, aiming to reveal the scope of crisis management plans and the range of their applicability, had been prepared from a close study of national and international regulations and requirements.

Table 1: Actions Taken by Ports for Crisis Management

Variable	n	%	Variable	n	%
Crisis Management Plan			*Communication Means**		
Yes	13	100	VHF	13	100
No	-	0	MF-HF Radio	11	84.6
Crisis Management Team			Mobile Phone	2	15.3
Authority from Port Operation	6	46.1	Others	1	7.7
Department					
Technical Advisor	5	38.4	*Rehabilitation Schedules*		
Director	2	15.3	Yes	-	-
Executive from Personnel	2	15.3	No	13	100
Department					
General Director	2	15.3	*Incentives*		
Others	3	23.0	Yes	-	-
Allocation of Stand-By Staff			No	13	100
Yes	6	46.1	*Recent Crisis*		
No	7	53.9	Fire	6	60
Scenarios in Management			Natural Disasters	4	40
Plan					
Yes	9	69.2	Job Accident	2	20
No	4	30.8	*Consequences of Recent*		
			Crises		
Fields Involved in			Full Success	8	88.8
*Management Plan**					
Fire	8	61.5	Partial Success	1	11.2
Job Accidents	8	61.5	*Reason for the Failure*		
Natural Disasters	3	23.0	Shortage of Equipment	1	50
Wars	2	15.3	Inadequate Knowledge	1	50
Strikes	1	7.7	*Correcting Measures*		
Others	2	15.3	Taken, Planned, Applied	5	83.3
Cure Seeking Means in Plans			Planned, Not Applied	1	16.7
Practising at least once a year	5	38.4			
Practising more than once a year	3	23.0			
Periodical meetings	3	23.0			
Others	2	15.3			

* The total exceeds % 100 due to multiple responses.

Shortly before faxing the questionnaires to the ports, there had been two highly destructive earthquakes in northwestern Turkey, where most of the prevailing ports are located. Thus, no proper contact was possible with some of the leading ports. As a consequence, out of 20 ports reached, only 13 responded.

In Table 1 the contents of this questionnaire and the responses for each item can be seen.

Evaluation of the Questionnaire

The responses received to the questionnaire seem to indicate that

- All the ports in question have crisis management plans, which is understandable as national regulations require this;
- Technical advisors and operation departments are dominant in crisis management;
- Most of the crisis management plans do not designate any stand-by staff. This might be considered negligent;
- Most of the scenarios are based on fires and accidents, which are clearly the prevailing crises faced;
- The responses given to 'cure-seeking means' could reveal that the ports in question intend to do the minimum towards meeting formal/national regulation requirements;
- The most common communication means in crises are VHF and HF-MF radios;
- Fire is regarded the most dangerous crisis;
- Life-saving is the dominant priority;
- Physical location and the number of employees seem to be the most effective factors in shaping crisis management plans;
- Fire-fighting departments, local government and coastguards seem to be the most important institutions with which to establish cooperation;
- Most of the ports in question have limited access to vital measuring instruments;
- No rehabilitation and/or compensation and/or incentives are considered in crisis management plans;
- Most of the crises experienced in Turkish ports seem to have been fires and the relevant ports seem to have had success in dealing with them.

Conclusions and Recommendations

This study aimed to search for an efficient means of minimising damage at ports in general and Turkish ports in particular, stemming from crises. The study considered the best possible proactive and reactive means of combating probable crises. Considering the technology and capital intensive infrastructures of ports, the vital importance of establishing effective and practicable crisis management plans was clear. In addition, the study proposed a model for developing efficient crisis management plans which constitute three basic steps - preparedness response and restoration corresponding to the pre-crisis, crisis, and post-crisis periods respectively. While considering the proactive and reactive actions to be taken in the case of crises in general, the study has tried to outline the prevailing ports in Turkey in terms of their present approaches towards crises. Considering national and international regulations which lay out actions to be taken against crises as well as analysing the responses from the ports, the study has aimed to reveal inadequacies in the approaches of the leading Turkish ports. An overall evaluation of the responses to the questionnaire revealed that Turkish ports in general have focussed on firefighting but ignored the other sources of crises, mainly the natural ones such as earthquakes and floods, despite the fact that such natural disasters have recently been experienced in Turkey quite often causing considerable damage. The study has also revealed that Turkish ports should place more focus on training, motivating and practicing, proactive measures aiming to lessen the damages likely to be brought about by quite common crises stemming from human-error, shortage of technology and ill-designed coordination and co-operation.

References

Kisi, H. (2000) Contingency Planning in Marina Management, *Proceedings of the First International Joint Symposium on Business Administration*, Gokceada, Canakkale, 1-3 June 2000.

Ross, J.E and Murdick R. (1973) *Management Update: the Answer to Obsolesence*, American Management Association.

Summers J. (1977) Management by Crisis, *Public Personnel Management*, May-June.

Tüz V.M. (1996) *Kriz Döneminde İşletme Yönetimi* (*Business Management during Crisis*), Bursa: Ekin Print.

Weitzel W. and Johsson E. (1989) Decline in Organisations: a Literature Integration and Extension, *Administrative Science Quarterly*, 34.

http://www.ege.com/2000

http://www.env.gov.bc.ca

7 Shipbuilding Markets and the Impact of Technology as a Macro Environmental Factor

A. GÜLDEM CERIT
SCHOOL OF MARITIME BUSINESS AND MANAGEMENT,
DOKUZ EYLÜL UNIVERSITY, IZMIR
OSMAN KAMIL SAĞ
MARITIME FACULTY ISTANBUL, TECHNICAL UNIVERSITY

Introduction

Technology provides a macro external environment for the shipbuilding industry and technological developments have considerable effects on the structure of the shipbuilding markets. New technologies that provide superior value in satisfying needs of customers stimulate investment and economic activity. Shipbuilding markets are affected by the accelerating pace of technological change and unlimited opportunities of innovation. Technological changes and safety requirements result in increased regulations and shipbuilding companies are responsible for the application of these regulations. These requirements yield to more investment in research and development activities. In this manner in terms of shipbuilding companies, technological changes have an impact on their business functions. Technology affects the design, construction, engine room capabilities and the internal decoration of the ships. This study aims from a Turkish perspective, to analyse the impact of technology on the global shipbuilding markets concentrating on its effects on the nature of the market. Technology is a tool for shipbuilding companies to compete effectively in the global markets and analysis of the effects of technology on the markets will help the industry develop new marketing strategies.

Economics and Technology

Technology has long been discussed in the scope of both economics and business disciplines. Contemporary theories on competition have paid considerable attention to the affects of technological developments.

Economists define the central economic tasks of every society as choices among an economy's inputs (factors of production) and outputs (goods or services) and an economy uses its existing technology to combine inputs to produce outputs (Samuelson, 1992). Technological change is explained as referring to changes in technology, invention of new products, improvement in old products, or changes in the processes for producing goods and services (Samuelson, 1992). Technological advance may result either by learning and adapting from others or through investments in research and development (R&D) (Nelson, 1994; Arrow, 1994; Freeman, 1994).

Competition and Technology

Porter describes technological change as one of the principal drivers of competition and supports the idea that 'technological change is not important for its own sake but is important if it affects competitive advantage and industry structure' (Porter, 1985). Technological change can affect competition through its impact on creation of value through any activity (Porter, 1985). Technology affects competitive advantage if it has a significant role in determining relative cost position or differentiation (Porter, 1985; Ramaswamy *et al,* 1994; Beard and Easingwood, 1992). Advantages created through technological changes are scale, timing, or interrelationships.

In international markets, new technologies are at the front of shifting competitive advantage by creating new possibilities for the design of a product, the way it is marketed, produced or delivered (Porter, 1990; Craig, and Douglas, 1996).

Strategic Management and Technology

Firms obtain sustained competitive advantage by implementing strategies that make use of their internal strengths through responding to environmental opportunities while neutralising external threats and avoiding internal weaknesses (Barney, 1991; Cerit *et al,* 1997).

Global strategies are affected by both the macro and micro external environmental factors and the internal environmental factors of businesses. The external macro environment consists of the demographic, economic, technological, natural, social/cultural and political/legal forces. External micro environment actors are grouped as customers, competitors, distribution channels and suppliers. Internal environmental factors are related to the business functions of marketing, finance, manufacturing and organisation (Terpstra and Ravi, 1994; Keegan and Green, 1997; Kotler, 1997; Cerit, 2000).

Strategic Market Planning and Technology

The strategic market planning process involves analysis of the situation, and the formulation, evaluation, selection, implementation and control of marketing strategies (Kerin *et al,* 1990). Situation analysis covers analysis of the environmental forces where techno-environmental conditions play an important role. Technology affects all business functions, information and communication system technologies constituting a global network (Figure1).

From a marketing point of view, technological change affects any firm competing in a certain industry. Previous studies in the marketing field analysing the effects of technology have focused upon:

- Competitive advantages produced by innovations in new-product development and new process development (Kerin *et al,* 1992; Song and Perry, 1997; Bello and Gilliland, 1997);
- Licensing in product development strategies (Kotabe, 1990; Kotabe *et al,* 1996);
- Competitive effects on technology diffusion and the results faced in the marketing field (Robertson and Gatignon, 1986; Gatignon and Robertson, 1989).

Marketing's role in the product innovation process and the effects in research and devlopment activities have produced a major area of interest for marketing scientists (Hutt and Speh, 1984; More, 1984; Gupta *et al,* 1986).

Figure 1: Technology as a Macro Environmental Factor for Business and its Impact upon Corresponding Functional Technologies

ORGANISATION
• Office Technology
• Administrative Technology
• Training and Human Resources Management Technology

MARKETING
• Software Technology
• Logistics and Transportation Technology
• Promotion, Sales/Post Sales Technology

FINANCE
• Office Technology
• Software Technology
• Financial Management Technology

MANUFACTURING
• Materials and Logistics Technology
• Process and Control Technology
• Product and R&D Technology

For safety purposes, technological change is associated with increased regulation. This surrounds industries and firms with another competitive environment forcing them to manufacture and market in accordance with regulations (Kotler, 1997).

Shipbuilding Industry and Technology

The shipbuilding industry is a sub-sector of manufacturing industry, serving the transportation industry, and shipbuilding markets possess the characteristics of industrial markets (Haas, 1989; Cerit, 1997a). Technology has an important impact on industrial markets where specifications and professional purchasing play an important role. The total worldwide shipbuilding capacity is estimated to be far in excess of global requirements for the beginning of the twenty first century. However both capacity increases in recent years and developments in shipyard technologies have produced competitive difficulties which have resulted in severe price cuts in the industry.

Recent studies examining approaches to the competitive drawbacks of shipbuilding businesses have focused mainly on:

- Strategic responses of the shipbuilding industry to the competitive environment (Hughes, 1989; Crezo, 1996; Slack *et al*, 1996; Marchese, 1997);
- Analysis of the shipbuilding business in the functional areas of finance (Cerit and Caki, 1996), cost and price relations (Cerit, 1997a), market orientation (Cerit, 1997b) and innovation (Johannessen *et al*, 1993);
- Shipyard development and productivity strategies (Birmingham *et al*, 1997; Sladoljev, 1998; McAlear, 1998; Bendall and Stent, 1996; Bruce and Garrard, 1999).

Businesses perform in an environment of technology and technological change affects competition. This case applies to the shipbuilding industry as well.

Objectives

The shipbuilding industry is one of the most concentrated industries where specifications and regulations are internationally approved. 'Shipbuilding contracts' which are applicable worldwide through the standard contracts of international or regional shipbuilders' associations, cover all the aspects of the shipbuilding schedule, where technology dictating specifications and regulations, plays the leading role. Shipbuilding markets are affected internationally by this leading role and this study aims to present an approach to the analysis of the impact of technology on the shipbuilding markets as a macro environmental factor.

Considering the effects on the business functions, technology is an important determinant for the functional areas of the shipbuilding industry where manufacturing constitutes the main interest area. The design, construction, engine room capabilities and the internal decoration of the ships are a result of technological changes.

Internationally regulated shipbuilding standards and competition in the context of the development of new technologies, keeps the shipbuilding industry in a challenging environment within which the industry has to survive and succeed in the marketplace.

Technological Impacts upon Shipbuilding Markets

The shipbuilding industry is at the service of the transportation industry and the demand for the shipbuilding industry is derived from the demand for the shipping industry (Cerit, 1997a). The decrease in the growth rates of world exports after 1995 caused a corresponding decrease in the growth of world seaborne trade and although a rise was observed in 1997 (UNCTAD, 1998), the developments in 1998, especially the severe economic conditions in the Far East, have affected growth inversely. Moreover the trends toward differing demands in the shipping industry have caused changes in the structure of the world merchant fleet. The shipbuilding industry, already facing an overcapacity environment, has also faced the challenges produced by the world economy. In an attempt to create competitive advantage, technology has been an important factor serving the industry.

Maritime Transport and Technological Trends

World seaborne trade passed the 5 billion tons mark in 1998 and the total world fleet amounted to 775.9 million dwt at the end of 1997 (UNCTAD, 1998). Developments in containerisation have produced new challenges for the shipping industry. Regarding the 1997 figures, containerships accounted for 8.1% of all the vessels in the world fleet in dwt terms. At the beginning of 1999 the total capacity of the container fleet was 6.77 million TEU, of which cellular vessels represented 70% (4.2 million TEU) (IAPH, 1999). During the 1993-1997 period, the container fleet faced a volume increase of 74% in a total of five years (UNCTAD, 1995; 1998). Technological developments aimed at reducing operating costs have been primarily reflected in increased vessel sizes. Due to increase in ship sizes, developments in total world TEU capacity since 1995 show a 3.6% increase in carrying capacity per ship (UNCTAD, 1998).

Technological developments have been effective in expanding ship sizes in containerisation. The first generation of container ships were tankers or general cargo vessels which were rapidly succeeded by second and third generation ships that differed in speed, motive power and capacity. By the 1980s a threshold ship size was reached, a limit determined by the dimensions of the Panama Canal; Panamax ships, with a capacity of about 3,300 TEU. In the 1980s the first steps were taken to build post-Panamax liners and the trend is towards even larger vessels (Slack *et al*, 1996). It is expected that the first 8,000 TEU vessels will start operations at the beginning of the 21st century (Payer, 1999). Another prediction is that by the year 2010, several 15,000 TEU vessels will be in operation on the biggest maritime routes.

These large ships are preferred for economies of scale on certain high density trade routes such as the United States West Coast to Asia. The expansion in containerisation is associated with investments to increase space requirements at ports and each are correlated with technical, organisational and territorial innovations.

Information technology requirements such as Electronic Data Interchange (EDI) play another important role in investments in technology and the shipping industry faces competitive pressures due to the necessity of these investments.

Another trend in the industry is towards safety and environmental regulations covering all aspects of the maritime transportation industry. The requirements for double-hulled tankers have been a major area of technological investment for oil tankers (Brown and Savage, 1996). Other internationally applied regulations leading to technological developments in both operations and training cover the SOLAS (Safety of Life at Sea) and ISM (International Safety Management) Code requirements, GMDSS (Global Maritime Distress and Safety System) equipment and training requirements, PSC (Port State Control) training and application requirements, and STCW (International Convention on Standards of Training, Certification and Watchkeeping for Seafarers) requirements for training of seafarers, all of which are the responsibilities of government bodies (IMO, 1997; ICS and ISF, 1996; IMO, 1996; Sag, 1997a, 1997b).

Need for new technologies stemming either from the requirements of the world fleet or from the increased number of regulations has directly been reflected in the shipbuilding industry.

Shipbuilding Industry and Technological Developments

A decrease in the growth rate of world merchandise exports and world seaborne trade has affected the shipbuilding industry from the mid 1990s (OECD, 1997) and a considerably smaller number of newbuilding contracts was placed in 1996. However starting from 1997, the industry recovered mainly due to the demand for tankers, bulk carriers and containerships. The substantial ordering of large containerships mainly increased in the feeder service sector with Panamax and post-Panamax size vessels. The largest increase was in the VLCC sector and the tanker market is expected to be the major source of newbuilding activity partly due to the effect of the increased number and severity of regulations, and partly due to increasing demand for larger tanker capacities. The technologies for newbuildings are affected by the nature of this demand.

World tonnage on order at the end of 1997 reached 81.2 million dwt showing an increase from the previous years. Dry bulk carriers, oil tankers and containerships share the first three rows of world tonnage on order (UNCTAD, 1998). The leading shipbuilding countries - Japan and South Korea - accounted for a total of 69% of world newbuilding deliveries in 1998 (Lloyd's Register, 1999).

The overcapacity in the shipbuilding industry has been reflected in a decrease in shipbuilding prices, particularly for dry bulk carriers as a result of the deterioration in dry bulk freight rates.

After the year 2000 it is believed that severe competition will take place in the industry due to substantial overcapacity. In order to cope with the changes in market conditions, increasing the capacity for creative technology and creating new demand in fields other than those in conventional shipbuilding are proposed (Crowley, 1997; Matuzawa, 1997; Lanz, 1999).

Impact of Technology on Differentiation and Cost Structures

Technology developments aim to gain competitive advantage through either cost leadership or differentiation. Changes in materials and logistics technology, process and control technology, product and research and development technology are the tools to achieve this objective.

From a materials point of view, introduction of high tensile steel has effected remarkable reductions in vessel weight. The paint industry is another area where technological developments have been introduced through materials technology.

Ship size, fuel consumption, speed, increased cargo handling productivity and the related regulations are the main variables that are to be optimised as techno-economic objectives (Hughes, 1989; Bendall and Stent, 1996). Without such optimisations, innovations would not be effective. The shipbuilding process covers analysis and applications of the properties of the end product - the ship.

Most types of vessels have experienced a steady increase in size aiming to realise economies of scale; however the larger the units of ships, the higher the fuel consumption and the less the speed. The speed of the ships used to be a useful regulating factor in times of increased market activity; however high fuel costs have forced cost effective developments in the industry to reduce fuel consumption. With regard to high speed technology, studies have concentrated both on economics and on proof of higher reliability (Farris II and Welch, 1998). Developments in the main engine propulsion systems have provided thermal efficiencies and hydrodynamic performance improvement. Previous research on vessel performance has concluded that a Panamax vessel has 47% improved techno-economic performance over its 1977 forerunner (Hughes, 1989).

Technology also offers significant potential for improvement in 'asset utilisation' for liner shipping such as bar-coding and computer tracking and for capacity management (Brooks, 1993). Improvements in the shipbuilding industry have introduced new electronic and satellite technologies as well, ending with reduction in the level of manning of vessels.

When negotiating a shipbuilding contract both the market conditions and regulations play important roles. Fuel consumption, speed and deadweight are the main techno-economic parameters of a shipbuilding contract. All the varied machinery and equipment stated in a shipbuilding contract must meet all regulations, safety requirements being the primary concern. SOLAS, MARPOL and OPA 90 requirements for safety of life at sea and prevention of marine pollution present shipbuilders with important responsibilities all of which are based on technological standards (Hughes, 1989). Safety regulations in construction and equipment of ships will produce additional costs which are to be compensated by growth in productivity of the new generation of vessels (Kubicki, 1997).

Modern ship construction methods make use of computer assisted design and manufacturing techniques (CAD/CAM) to enhance structural strength with the minimum amount of steelwork and substantial savings in design office manpower thus resulting in a decrease in costs (Crezo and Sanchez-Jauregui, 1996; Birmingham *et al*, 1997; Beck and Lord, 1998). The shipbuilding industry is one of the most labour intensive industries and to reduce man-hours, production line fabrication methods are applied.

Differentiation and prevention of additional costs due to low quality have moved shipyards towards an environment of quality assurance which is supported by several of the classification societies. An effective quality assurance system can help to make a success of the construction process.

The development of new technologies is led by those shipyards and shipbuilding nations, which have the highest production and consequently highest labour utilisation and highest costs. Another criterion is access to financial resources for research, either from private sources or from governments (Bruce and Garrard, 1999).

Shipbuilding Markets and Technology

Despite the severe economic crisis in the Far East at the beginning of 1998, Japanese, South Korean and Chinese shipyards have received substantial new orders proving that capacities will be utilised until the second half of the year 2000. However after the estimated fall in demand for tankers after 2000, it is believed that overcapacity problem will emerge in Japan since the bulk of tanker replacement should be completed by 2001. Japan's shipyards are believed to have cut and oversimplified their design departments in search of a decrease in manufacturing expenses (Seatrade Review, 1998).

The economic crisis did not change the market potential substantially for South Korea. Having the cheapest workforce and given expectations about tanker replacement and offshore projects over the next few years, Korea has an optimistic view of the near future (Middleton, 1998).

China has risen to become the world's third highest recipient of orders (in grt terms) for new vessels and accounts for 5-6% of newbuildings worldwide depending on recent technological developments and the cheap workforce in the industry. China stands ahead of Germany and behind Japan and South Korea and 85% of ships built by the Chinese shipyards were exported in 1997 (Stormont, 1997).

Far Eastern yards concentrate on building VLCCs, large containerships and 30,000 ton-plus bulk carriers while European yards dominate value added contracts for the cruise and offshore industries. European shipbuilders selected a strategy of specialisation and this has kept most of the yards out of difficulty (Anon. [1], 1998). Germany, being the fourth largest shipbuilding nation, seeks to diversify into higher value-added work (Seatrade Review, 1997).

The leading nations in the shipbuilding industry concentrate on different aspects of technological developments for quality assurance and research and development to attain market differentiation.

Conclusions

The effects of technology on the competitive advantage of businesses and the corresponding impact on business functions have long provided a medium of research for the business discipline.

The shipbuilding industry, where demand is derived from the shipping industry, is a unique application area due to the highly internationalised and regulated nature of the sector. Internationally developed specifications, standards and regulations based on technological requirements provide the members of the industry with related challenges.

Technology as a macro environmental factor for the shipbuilding markets, needs to be analysed empirically in depth with respect to competitive measures. Further studies may cover the aims of defining and measuring the technological variables and parameters affecting the competitive position of the industry and their individual impact on the individual businesses may be assessed.

Building effective marketing strategies for the shipbuilding industry, which is of a highly technological nature, requires a thorough analysis of the technological variables.

References

Anon. [1] (1998) Taking Nothing for Granted, *Seatrade Review*, March, 59.

Anon. [2] (1998) European Shipbuilding - Specialisation is Saving Grace, *Lloyd's Ship Manager*, June, 37.

Anon. [3] (1997) Heading for Rationalisation, *Seatrade Review*, September, 59.

Arrow, K. J. (1994) The Production and Distribution of Knowledge, in Silverberg, G. and Soete, L. (eds) *The Economics of Growth and Technical Change*, Edward Elgar Publishing Limited.

Barney, J. (1991) Firm Resources and Sustained Competitive Advantage, *Journal of Management*, 17 (1), 99-120.

Beard, C. and Easingwood, C. (1992) Sources of Competitive Advantage in the Marketing of Technology Intensive Products and Processes, *European Journal of Marketing*, 26/12/1992, 5-18.

Beck, D. and Lord, J. (1998) Design and Production of ANZAC Frigates for the RAN and RNZN: Progress Towards International Competitiveness, *Journal of Ship Production*, 14, 2, 85-109.

Bello, D.C. and Gilliland D.I. (1997) The Effect of Output Controls, Process Controls and Flexibility on Export Channel Performance, *Journal of Marketing*, 61, 22-38.

Bendall, H.B. and Stent, A.F. (1996) Hatchcoverless Container Ships: Productivity Gains from a New Technology, *Maritime Policy and Management*, 23, 2, 187-199.

Birmingham, R., Hall, S. and Kattan, R. (1997) Shipyard Technology Development Strategies, *Journal of Ship Production*, 13, 4, 290-300.

Brooks, M. (1993) International Competitiveness - Assessing and Exploring Competitive Advantage by Ocean Container Carriers, *The Logistics and Transportation Review*, 29, 4, 275-293.

Brown, R.S. and Savage, I. (1996) The Economics of Double-Hulled Tankers, *Maritime Policy and Management*, 23, 2, 167-175.

Bruce, G. J. and Garrard, I. (1999) *The Business of Shipbuilding*, London: Lloyd's of London Press Limited.

Cerit, A. G. and Caki, S. (1996) Financial Prospects for the Turkish Shipbuilding Industry, *Proceedings of the First International Conference on Marine Industry*, Varna, 2-7 June. Volume 1, 181-192.

Cerit, A. G. and Caki, S. (1997a) Price Element and the Shipbuilding Industry: an Analogical Approach to the Relations of Labor/Material Costs, *International Conference on Ship and Marine Research, NAV' 97*, Sorrento, 18-21 March, University of Naples Federico II, 1, 15-1, 29.

Cerit, A. G. and Caki, S. (1997b) Analysis of Market Orientation in a Concentrated Business Market: Shipbuilding Industry in Turkey, *Proceedings of the Eighth Congress of the International Maritime Association of Mediterranean*, Istanbul, 2-9 November, Istanbul Technical University, Volume II, 10.1-1-10.1-11.

Cerit, A. G., Caki, S., Kisi H., Tuna, O. and Saatcioglu, O. (1997) Gümrük Birliği Sürecinde Ege Bölgesi Sanayiine Rekabet Gücü Açisindan Stratejik Bir Yaklasim, *3. Verimlilik Kongresi*, 14-16 Mayis. Ankara: MPM Publication, 177-196.

Cerit, A. G. and Caki, S. (2000) Maritime Transport as an Area of Competitive Advantage in International Marketing, *International Journal of Maritime Economics*, II, 1, 49-67.

Craig, S. and Douglas, S.P. (1996) Responding to the Challenges of Global Markets, *The Columbia Journal of World Business*, Winter, 6-18.

Crezo, J. L. and Sanchez-Jauregui, A. (1996) Spanish Shipbuilding: Restructuring Process and Technological Updating from 1985 to 1995, *Proceedings of the First International Conference on Marine Industry*, Varna, 2-7 June, Volume 1, 169-180.

Crowley, E. (1997) World Shipbuilding and Shiprepair - an Overview, *BIMCO Review 1997*, London: BIMCO Publication.

Farris II, M.T. and Welch, D. (1998) High Speed Ship Technology, *Transportation Journal*, 38, 1, 5-14.

Freeman, C. (1994) Technological Revolutions and Catching Up, in Faberberg, J., Verspagen, B. and von Tunzelmann, N. (eds) *The Dynamics of Technology, Trade and Growth*, Edward Elgar Publishing Limited; Aldershot.

Gatignon, H. and Robertson, T.S. (1989) Technology Diffusion: an Empirical Test of Competitive Effects, *Journal of Marketing*, 53, 35-49.

Gupta, A.K., Raj, S.P. and Wilemon, D. (1986) A Model for Studying R&D - Marketing Interface in the Product Innovation Process, *Journal of Marketing*, 50, 7-17.

Haas, R.W. (1989) *Industrial Marketing Management*, Boston: PWS-KENT Publishing Company.

Hughes, C. N. (1989) *Shipping: a Technoeconomic Approach*, London: Lloyd's of London Press Ltd.

Hutt, M.D. and Speh, T.W. (1984) The Marketing Strategy Center: Diagnosing the Industrial Marketer's Interdisciplinary Role, *Journal of Marketing*, 48, 53-61.

IAPH - International Association of Ports and Harbors Committee on Ship Trends (1999) *Biennial Report on Ship Trends - 1999*, Tokyo. Unpublished Report.

ICS ve ISF - International Chamber of Shipping and International Shipping Federation (1996) *Guidelines on the Application of the IMO International Safety Management (ISM) Code*, London: ICS and ISF Publication.

IMO - International Maritime Organisation (1996) *STCW Convention Resolutions of the 1995 Conference, STCW Code*, London: IMO Publication, IMO-938E.

IMO - International Maritime Organisation (1997) *SOLAS (Consolidated edition)*, London: IMO Publication, IMO-110E.

Johannessen, J., Olaisen, J. and Hauan, A. (1993) The Challenge of Innovation in a Norwegian Shipyard Facing the Russian Market, *European Journal of Marketing*, 27, 3, 23-38.

Keegan, W. J. and M. C. Green (1997) *Principles of Global Marketing*, New Jersey: Prentice Hall.

Kerin, R.A., Mahajan, V. and Varadarajan, P.R. (1990) *Strategic Market Planning*, Boston: Allyn and Bacon.

Kerin, R.A., Varadarajan, R. and Peterson, R.A. (1992) First-Mover Advantage: a Synthesis, Conceptual Framework, and Research Propositions, *Journal of Marketing*, 56, 33-52.

Kotabe, M. (1990) Corporate Product Policy and Innovative Behavior of European and Japanese Multinationals: an Empirical Investigation, *Journal of Marketing*, 54, 19-23.

Kotabe, M., Sahay, A. and Aulakh, P.S. (1996) Emerging Role of Technology Licensing in the Development of Global Product Strategy: Conceptual Framework and Research Propositions, *Journal of Marketing*, 60, 73-88.

Kotler, P. (1997) *Marketing Management: Analysis, Planning, Implementation and Control*, New Jersey: Prentice Hall.

Kubicki, J. (1997) Economic Aspects of Safety in Construction, Equipment, Manning and Operation of the Vessels, *Proceedings of the Eighth Congress of the International Maritime Association of Mediterranean*, Istanbul, 2-9 November, Istanbul Technical University, Volume II, 10.1-16-10.1-20.

Lanz, R. (1999) Yards are Warned on Overcapacity, *BIMCO Review 1999*, London: BIMCO Publication, 258-260.

Lloyd's List (1998) 26th November, 1.

Lloyd's Register (1999) *World Fleet Statistics*, London: LR Publication, March.

Marchese, U. (1997) Intermodality and the Evolution of Competition in Shipping Markets, *International Conference on Ship and Marine Research, NAV' 97*, Sorrento, University of Naples Federico II, 1.3-1.14.

Matuzawa, S. (1997) Building for the Future, *BIMCO Review 1997*, London: BIMCO Publication, 389.

McAlear, R. (1998) Shipyard Modernisation - a Shipbuilder's Experience, *Journal of Ship Production*, 14,1, 1-9.

Middleton, I. (1998) Building to Bloom After a Dark Winter? *Seatrade Review*, May, 6.

More, R.A. (1984) Timing of Market Research in New Industrial Product Situations, *Journal of Marketing*, 48, 84-94.

Nelson, R.R. (1994) What Has Been the Matter with the Neoclassical Theory?, in Silverberg, G. and Soete, L. (eds) *The Economics of Growth and Technical Change*, Edward Elgar Publishing Limited: Aldershot.

OECD (1997) *Report on Council Working Party on Shipbuilding - WP 6*. Paris: Unpublished Report.

Payer, H. G. (1999) Large Containerships - Where is the Limit? *BIMCO Review 1999*, BIMCO Publication: London, 269-272.

Porter, M.E. (1985) *Competitive Advantage*. New York: The Free Press.

Porter, M.E. (1990) *The Competitive Advantage of Nations*. New York: The Free Press.

Ramaswamy, V., Gatignon, H. and Reibstein, D.J. (1994) Competitive Marketing Behavior in Industrial Markets, *Journal of Marketing*, 58, 45-55.

Robertson, T.S. and Gatignon, H. (1986) Competitive Effects on Technology Diffusion, *Journal of Marketing*, 50, 1-12.

Roe, M.S. (1998) *Commercialisation in Central and East European Shipping*, AshgatePublishing: Aldershot.

Sag, O.K. (1997a) 1997 Yılında Genel Denizcilik Egitiminde Meydana Gelen Gelismeler, *Mersin Deniz Ticareti*, April, 18.

Sag, O.K. (1997b) 1997 Yılında Genel Denizcilik Egitiminde Meydana Gelen Gelismeler, *Dünya, BIMCO Ozel Ek*, June 3.

Samuelson, P.A. and Nordhaus, W.A. (1992) *Economics*, New York: McGraw-Hill, Inc.

Slack, B., Comtois, C. and Sletmo, G. (1996) Shipping Lines as Agents of Change in the Port Industry, *Maritime Policy and Management*, 23, 3, 289-300.

Sladoljev, Z. (1998) Countries in Transition: Some of Croatia's Aims in Shipyard Productivity Achievement, *Journal of Ship Production*, 14, 1, 10-14.

Song, X.M. and Perry, M.E. (1997) A Cross-National Comparative Study of New Product Development Processes: Japan and the United States, *Journal of Marketing*, 61, 1-18.

Stormont, D. (1997) Aiming for 10% of World Market, *Seatrade Review*, 65.

Terpstra, V. and Ravi, S. (1994) *International Marketing*, Orlando: Dryden Press.

UNCTAD (1995) *Review of Maritime Transport 1995*, Geneva: United Nations Publication.

UNCTAD (1998) *Review of Maritime Transport 1998*, Geneva: United Nations Publication.

Yercan, F. and Roe, M.S. (1999) *Shipping in Turkey*, Ashgate Publishing: Aldershot.

8 Caspian Oil Exports and Their Impact Upon the Tanker Fleet

BENGI SELEN YÜCEER AND A. GÜLDEM CERIT
SCHOOL OF MARITIME BUSINESS AND MANAGEMENT,
DOKUZ EYLUL UNIVERSITY, IZMIR

Introduction

Oil has never lost its reputation as the major source of energy since 1945. While scientists try to find new types of energy, oil today continues to rule the world even more strongly than ever. Industry still depends on oil in many ways as it provides a larger share of world energy consumption than any other source, at 39% of the total in 1997.

The subject of this article, the Caspian Region, is also playing its part in the oil game. Many companies from western nations have invested in Caspian Oil. Amoco and British Petroleum, Chevron, the Atlantic Richfield Company and the Russian oil giant Lukoil are some of the major players in the region. Apart from the multi-national oil exploration companies, there are construction enterprises waiting to build pipelines, compressor stations, petrochemical plants, and ports and railroads, to realise the so called 'silk road of the 21st century'. The scene would be incomplete without mentioning neighbouring states that struggle to attract the silk road to their territory, such as Turkey, Russia, Iran, Afghanistan, Pakistan and even China.

This article will mainly focus on Caspian oil output and Turkey's projected key role in the exportation and the impact on tanker demand. Oil issues cannot be treated separately from economics or politics, and any work that focuses only on one side, may be considered incomplete. However in this article, the authors will not concentrate upon the political issues, which remain an important concern nevertheless.

The Caspian Region, Oil Reserves, Oil Producing Countries and Export Routes

The Caspian Region is a geographically landlocked area located in west Asia and includes Georgia, Azerbaijan, Kazakhstan, Turkmenistan, Uzbekistan, Iran and the Russian Federation.

Although most of the oil and gas reserves remain undeveloped and many areas still remain unexplored, the Caspian Basin is considered only second to the Middle East and larger than Alaska or the North Sea, with 32.5 billion barrels (bbl) proven and 163 bbl possible oil reserves (US Energy Information Administration, 2000a).

Table 1: Caspian Basin Oil Reserves (billion barrels)

	Proven oil reserves	Estimated oil reserves	Total
Azerbaijan	3.6-12.5	27	**31-40**
Iran*	0.1	12	**12**
Kazakhstan	10.0-17.6	85	**95-103**
Russia*	0.3	5	**5**
Turkmenistan	1.7	32	**34**
Uzbekistan	0.3	1	**1**
TOTAL	**16-32.5**	**163**	**179-195**

* Only Caspian Region reserves are included.
Source: US Energy Information Administration (2000) *Caspian Sea Region.*

The three major oil producing countries in the region, Turkmenistan, Azerbaijan and Kazakhstan, have both on and offshore wells of 31.8 bbl proven oil reserves. The total production of these three countries could reach over 4 million barrels per day (b/d) by the end of the next decade, one fourth of which is expected to be exported out of the region. By 2020, it is estimated that production and exports could increase by another 2 million b/d (U.S. Energy Information Administration, 2000a).

Meanwhile, Russia has discovered an offshore oil reserve as a recent development in the Caspian Basin. As the exploration activities continue, findings confirm or exceed the original estimations.

Major Oil Producing Countries in the Caspian Region and Alternative Export Routes

Transportation of oil is still an important concern for the region's developing countries. Although the market is ready to consume whatever will come from the region, rich reserves cannot be transformed into cash without viable modes of transportation. The Caspian Region's high oil production rates (Table 2) and its distance from the sea, automatically cancel out road, rail and sea transportation alternatives. For a landlocked region like the Caspian, the most rational solution seems to be pipeline transportation.

Table 2: Caspian Sea Region Oil Production and Exports (000 b/d)

Country	Production 1990	Net Exports 1990	Production 1997	Net Exports 1997	Possible Exports 2010
Azerbaijan	259.3	76.8	192.9	58.4	1,000-1,500
Kazakhstan	602.1	109.2	573.3	310.9	2,000
Turkmenistan	124.8	69.0	107.3	39.2	50
Uzbekistan	86.2	-168.1	182.4	23.8	50
Russia*	144.0	0.0	60.0	0.0	0.0
Total	1,216.4	86.9	1,115.9	428.7	3,100-3,600

* Includes the North Caucasus region bordering the Caspian Sea
Source: U.S. Energy Information Administration (2000) *Caspian Sea Region.*

Azerbaijan

Azerbaijan's proven oil reserves are 3.6-12.5 bbl as of January 2000 and crude oil production realised in 1999 was 255.000 b/d, approximately half of which was exported (U.S. Energy Information Administration, 2000a).

Several options for pipeline routes to carry oil to world markets were presented for consideration to the Azerbaijan government, including:

- Baku to Ceyhan (Turkey);
- Baku to Supsa (Georgia);
- Baku to Novorosiisk (Russia).

Transporting oil from Baku (Azerbaijan), to the terminal at Ceyhan on the East Mediterranean coast of Turkey, is the Caspian oil export option that the U.S., Turkish, and Azerbaijani governments prefer. Although the Baku-Ceyhan Pipeline is considered expensive when compared to the other two, it is the preferred option as it passes through a more stable region near to the potential markets, and as such, it provides a cheaper and less risky sea transportation alternative without the dangerous passage through the Strait of Istanbul (Figure 1).

Kazakhstan

Kazakhstan is important for world energy markets with her significant oil and gas reserves as she is the second largest oil producer among former Soviet Republics after Russia, producing over half a million b/d (U.S. Energy Information Administration, 2000a).

Kazakhstan exports about 170,000 b/d of crude oil through the Russian pipeline system; by barge and rail to the Baltic; and by ship, pipeline, and rail to the Black Sea. Given adequate export outlets, this figure is estimated to reach peak production of 750,000 b/d by 2010 (U.S. Energy Information Administration, 2000a). Kazakh oil from the Tengiz field will be exported by the Caspian Pipeline Consortium (CPC) to world markets via a 900-mile oil export pipeline connecting to the Russian Black Sea port of Novorosiisk. Construction of the pipeline is underway and the pipeline is expected to be commissioned in 2001 but it will not reach full capacity of 1.34 million b/d until about 2015 (U.S. Energy Information Administration, 2000a).

As can be seen in Figure 1, several other routes were proposed to transport Kazakh oil towards markets in Asia:

- One proposed pipeline would bring Kazakh oil to outlets in Iran and the Persian Gulf. The governments of Kazakhstan and Iran agreed in November 1997 to resume oil exchange between the two countries.
- In addition, the proposed Central Asia Oil Pipeline would bring oil from Kazakhstan to Pakistan and the Arabian Sea.
- Kazakhstan is also considering the Chinese market, and in June 1997, the China National Petroleum Corporation signed an agreement under which China will invest $3.5 billion to build an 1,800-mile pipeline from Yangiz to Xinjiang China (U.S. Energy Information Administration, 2000a).

Turkmenistan

The other important oil producing country in the Caspian Region is Turkmenistan as she has 1.7 bbl proven oil reserves (Table 1). One of the main obstacles preventing the development of Turkmenistan's oil industry is the lack of export routes (U.S. Energy Information Administration, 1999).

Like other countries in the region, Turkmenistan is also seeking alternatives for the transportation of its oil. In March 1998, the U.K.'s Monument Oil Company reached an agreement with Iran's National Iranian Oil Company (NIOC) to provide oil from the offshore Burun field in western Turkmenistan to the northern border of Iran and exchange it for oil to be exported from the Persian Gulf (Figure 1).

The country is considered capable of much greater production once fully developed, if a permanent export route becomes available (U.S. Energy Information Administration, 1999).

Uzbekistan

Uzbekistan has 0.3 bbl proven oil reserves (Table 1). The state owned oil and gas company has projected that by 2010, crude oil output of Uzbekistan will reach 240,000 b/d (U.S. Energy Information Administration, 2000a).

The lack of export pipelines in the region adversely affects Uzbekistan as is the case of other Caspian Region countries. Without viable pipeline routes Uzbekistan will be unable to sell its oil on world markets. Uzbekistan's only current option for transporting oil is to reverse an existing crude oil pipeline that brings oil from Omsk, Russia to Uzbek refineries (U.S. Energy Information Administration, 2000a).

In order to overcome these problems, Uzbekistan has signed a Memorandum of Understanding (MOU) with Turkmenistan, Afghanistan and Pakistan to build the Central Asia Oil Pipeline (CAOP). If constructed, the CAOP would transport Central Asian oil via Afghanistan to a proposed new deepwater port at Gwadar on Pakistan's Arabian Sea coast.

Russian Federation

Russia is luckier with its transportation alternatives when compared to other Caspian Region countries. The majority of Russian oil is exported via terminals in the Baltic and Black Seas. The most important problem in Black Sea exports is the increasingly crowded Turkish Straits passage. Russia continues to campaign for the main export pipeline of the Azerbaijan International Oil Consortium (AIOC) in Azerbaijan to be routed through Russia to the port of Novorossiisk instead of to Ceyhan or to the Black Sea port of Supsa, Georgia, where a 100,000 b/d pipeline is already in operation (Figure 1).

Russia's Transneft Oil Transport Company announced in April 2000 the completion of the Baku-Novorossiisk pipeline bypassing Chechnya, despite a cost of $250-$300 million.

Baku-Ceyhan Pipeline Project

The Caspian Region has been at the centre of attraction with its vast resources of oil lying under its land for more than two centuries. It was Robert Nobel who first started to operate a modern oil well in Baku in 1873, and until the end of 19th century the Caspian Region provided nearly half of the world's oil needs (Pala and Engür, 1999). Today, the Caspian Region is struggling to revert to its past golden days.

The appetite-whetting phrase of 'rich oil reserves' would mean nothing unless the oil lying under the land has the chance to be marketed. Among the many solutions proposed for the transportation of Caspian oil to the world markets, the Baku-Ceyhan Pipeline Project received the highest support both politically and economically, from the interested nations in the region and from the USA.

Discussions that had been ongoing since 1992 concerning the Main Export Pipeline (MEP) issues, reached a conclusion on November 18, 1999 with the signing of a 'framework agreement' between Turkey, Azerbaijan and Georgia that favours construction of a 1,080 mile (1,730 kilometres) line from Baku, Azerbaijan, to Ceyhan on Turkey's Mediterranean coast. This project is aimed to carry Asian oil with minimum expense and with maximum security. All these agreements entered into force after they were signed by the Turkish Parliament on 23 June 2000. Turkey expects to earn 100 million US$ annually when (if) the pipeline is constructed.

The 1,080 mile pipeline from the Azeri capital Baku, through Georgia to the Port of Ceyhan on the Southeast coast of Turkey, is planned to carry 45 million tons of crude oil per year, 20 million tons/year of which will be from Kazakh fields, and 25 million tons from Azerbaijan fields. The construction of the pipeline is projected to begin in early 2001 for completion in 2004. Some 468 km of the pipeline will pass through Azerbaijan, while 225 km will be built in Georgia and the other 1,037 km will pass within Turkish territory until it reaches the Mediterranean (Figure 1). It was feared that the construction programme would be delayed due to several problems on Georgia's side, concerned with nationalist issues, environmental standards, security of the pipeline and transit charges. However, with the efforts of parties concerned with the Main Export Pipeline (MEP), all of these problems have been solved recently.

The Baku-Ceyhan pipeline project will strengthen the strategic position of Turkey, as the country will carry the most important oil route between oil producing Asia and demand centred Europe. Turkey will have the chance to house three of the five transportation modes; namely road, sea and pipeline transport, as a result of its geographical and infrastructural characteristics. Turkey forms a natural bridge between Asia and Europe with highways crossing the country from east to west and forms an optimal region for shipping activities with its 8,300 miles coastline. The scene for transportation will be completed after the construction of the Baku-Ceyhan Main Export Pipeline.

The Baku-Ceyhan Pipeline is planned to reach its full capacity of 45 million tons in the sixth year after its construction. The proposed flow of oil in the first year is about 11 million tons, this amount increasing to 18 and 25 million tons in the second and third years of the line's operation. The projected oil flow will be 32 million tons in the fourth year, and 39 million tons in the fifth (Table 3).

The Port of Ceyhan, where the oil carried by pipeline will be shipped to tankers, constitutes the most convenient port in the region for the transport of Caspian oil to world markets for a number of reasons. Firstly due to its proximity to Caspian and Middle East oil production areas, secondly the suitable climate and port conditions, and thirdly for it being a specialised port in handling crude oil. The Port of Ceyhan was used for oil shipments from Iraq for years; however oil shipments from this country have long been interrupted with the United Nations embargo on Iraq following the Gulf War. The region is expected to regain its lively economic atmosphere with the oil flow from the Baku-Ceyhan pipeline.

The Port of Ceyhan was built primarily for oil shipments. The Port has four terminals suitable for the berthing of tankers up to 300,000 dwt. Several projects are on the way for the expansion and renewal of the Port of Ceyhan. The renewed oil terminal is projected to be completed at the same time as the Baku-Ceyhan Pipeline in 2004.

Table 3: Oil Flow Projections of Baku-Ceyhan Pipeline (million tons)

1st Year	2nd Year	3rd Year	4th Year	5th Year	6th Year
11.5	18.0	25.0	32.0	39.0	45.0

Source: Information from officials of Republic of Turkey Ministry of Energy and Natural Resources, 2000.

World Oil Market and Impacts of Caspian Oil Exports Upon the World Tanker Fleet

In 1998, petroleum continued to be the world's most important primary energy source, accounting for 39.8% of world primary energy production (US Energy Information Administration, 2000b). Between 1970 and 1999, petroleum production increased 51%, rising from 48.9 to 74 million b/d. In the same period world oil demand increased 63.6%, from 46.8 million b/d to 74.5 million b/d (Table 5).

World Oil Demand

Between 1970 and 1997 oil consumption rose by a total of 26.2 million b/d, with an average annual increase of 1.7%. The oil market boomed in the 1990s, with demand fuelled by growing Asian countries and steady economic growth in the industrialised world. In 1997 oil demand grew by 2.8%, the highest rate of growth since 1988 and well above the 1.4% annual average for the ten year period (Peters, 2000). The worldwide crude oil consumption is estimated to be 93.5 million barrels in 2010 and 112.8 million barrels in 2020 (US Energy Information Administration, 2000b).

World Oil Supply

In 1998 Saudi Arabia, the United States and Russia were the three largest producers of petroleum, responsible for 31.9% of world output. Production from Iran and Mexico accounted for an additional 9.9% (US Energy Information Administration, 2000b). However, OPEC producers are expected to be the major beneficiaries of increased production requirements, but non-OPEC supply is expected to remain competitive, with major supplies especially coming from the Caspian Basin. However, the low price environment of 1998 and early 1999 did slow the pace of development in some production areas, especially the Caspian Basin region (McMahon, 2000).

Table 4: World Crude Oil Demand and Projections

	Million barrels/day	Billion tons/year
1997	73	3.4
2010	94	4.5
2020	111	5.3

Source: Pala, C. and Engur, E. (1999).

Table 5: World Oil Supply and Demand (000 barrels per day)

Year	Supply	Demand
1970	48,986	46,808
1980	64,152	63,067
1990	66,754	65,977
1997	73,650	73,001
1998	75,151	73,644
1999	74,005	75,581

Source. US. Energy Information Administration (2000b).

World Tanker Fleet: Current Potential and Future Projections

Together with the drilling of the first oil well in 1859, and rapid developments in the transportation industry, demand for oil continued to rise leading to the emergence of specialised vessels to carry this product in bulk. The first oil tankers were only 3,000-5,000 dwt and the largest tanker built in the 19th century was 12,800 dwt. The continuous rising demand for oil affected the size of oil tankers, which today has resulted in the creation of super tankers with sizes over 250,000 dwt.

Current Situation of the World Oil Tanker Fleet

At the end of 1998 the percentage of oil tankers in the world merchant fleet has reached 35.4, creating the largest group of merchant vessels worldwide.

Table 6: World Fleet Size by Vessel Types (end of 1998 - 000 dwt)

Vessel Types	dwt (000)	Percentage Shares
Oil Tankers	289,509	35.4
Bulk Carriers	281,012	34.9
General Cargo	101,259	12.8
Container Ships	61,147	7.8
Liquified Gas	16,471	2.1
Chemical Tankers	7,740	1.0
Miscellaneous	885	0.1
Ferries and	4,803	0.6
Others	41,392	5.2
TOTAL	788,725	100.0

Source: UNCTAD (1999).

The world oil tanker fleet is relatively old, as oil tankers over 15 years of age comprise 51.4% of the whole oil tanker market (Table 7). In the 1990s considerable new legislation has been introduced to overcome the marine pollution problem, including the International Convention for the Prevention of Pollution from Ships (MARPOL) from IMO and the Oil Pollution Act (OPA'90) of the United States. The new legislation will also affect tanker tonnage worldwide. For example OPA'90 will gradually bar single hull tankers from trading to the USA; on the other hand MARPOL mandates the retirement of all single hull tankers in international trade at 30 years of age. To trade beyond 25 years of age, pre-MARPOL tankers must retrofit protected spaces or make use of hydrostatically balanced loading in selected cargo tanks (Doll, 1999). The European Commission also proposed a package of legislation on maritime safety in the wake of the *Erika* oil tanker disaster. This package will include accelerated phasing out of single hull tankers and their replacement by double-hull vessels (Tutt, 2000). This is now likely to come about by 2015 following the decision of IMO to recommend a similar ruling.

At the beginning of 1999, the expected deliveries of new tankers were

35 VLCCs (with an additional 35 to be delivered in 2000), 20 Suezmax tankers and 50 Aframax tankers. As of August 1999, the tanker fleet in deadweight terms had grown about by 3% to 290 million dwt (McMahon, 1999). Although oversupply of oil tankers amounted to 20% of the oil tanker sector in 1999, due to growing concern about the environment and developments in liability issues, many old tankers were doomed to scrapping, while new tankers continued to be ordered.

Table 7: Age Distribution of the World Oil Tanker Fleet, 1998 (% dwt)

	0-4 years	5-9 years	10-14 years	15 plus	Average age 1998*
Oil Tanker	17.8	16.6	14.3	51.4	14.54

*To calculate the average age, it has been assumed that the ages of vessels are distributed evenly between the lower and upper limit of each age group. For the 15 years and over age group, the mid-point has been assumed to be 22 years.
Source: UNCTAD (1999).

Table 8: World Tonnage on Order (end of 1998 - 000 dwt)

	All Ships	Oil Tankers	Dry Bulk Carriers	General Cargo	Container Ships	Other Ships
DMEC*	25,515	12,666	3,703	2,589	3,516	3,041
MORC**	45,379	23,151	13,340	1,645	3,108	4,135
World Total	82,180	41,231	19,801	5,687	7,473	7,987

* Developed market-economy countries
** Major open-registry countries
Source: UNCTAD (1999).

Experts predict that the world oil tanker fleet could actually decline over the next two years as older vessels are scrapped after the *Erika* catastrophe.

In 1999, some 129 tankers of 14 million dwt were scrapped, a rise of 88.5 % on the year before. One estimate is that 40 vessels will be scrapped both this year and next, before falling to 30 vessels in 2002 and 10 vessels in 2005. In 2001, it is foreseen that the active VLCC fleet will fall from 434 to 419 vessels, reducing to 409 vessels in 2002. Thereafter, it is predicted that the fleet will expand to 414 vessels in 2003, 429 vessels in 2004, and 444 vessels in 2005 (Gray, 2000).

Current Potential of the Turkish Tanker Fleet

The Turkish merchant fleet is the 17th biggest fleet in the world (1999). In spite of a million dwt loss of its fleet due to the Asian crisis in the late 1990s, Turkey still preserves its place with a fleet of total 9,760,370 dwt (TCS, 1999).

Table 9: World Tonnage Sold for Breaking, 1993-1998 (000 ton/%)

	1993	1994	1995	1996	1997	1998
Tankers	10,665	13,102	10,877	6,550	3,578	7,426
	63.3	63.1	71.0	36.1	24.2	29.4
Combined Carriers	2,040	2,559	1,228	1,861	423	1,435
	12.1	12.3	8.0	10.3	2.9	5.7
Dry Bulk Carriers	2,645	3,829	2,135	7,632	8,161	12,847
	15.7	18.4	13.9	42.1	55.1	50.9
Others	1,502	1,282	1,081	2,092	2,646	3,533
	8.9	6.2	7.1	11.5	17.9	14.0
Total	16,852	20,722	15,321	18,135	14,808	25,241

Source: UNCTAD (1999).

Amongst the vessel types, bulkers form the biggest group with a percentage of 64.1 and a total of 6,256,314 dwt. Dry cargo ships and oil tankers rank after the bulkers; dry cargo ships are the second largest group with a percentage of 14.5 and oil tankers rank as the third largest group consisting of 803,121 dwt and with a percentage of 8.2 among the whole merchant fleet (TCS, 1999).

Although the number of tankers utilising the Strait of Istanbul during 1999 was 5,504, Turkey seems unable to evaluate the potential market transiting her waters. As of December 1998, Turkish flagged oil tankers over 150 grt are 98 in number, seven of which are owned by public companies. There are no ULCC or VLCC ships flying the Turkish flag (TCS, 1999).

According to statistics prepared by the Turkish Republic, Prime Ministry Undersecretariat of Maritime Affairs, foreign flagged tankers exceeded the number of Turkish flagged vessels in 1999 handling oil in Turkish Ports, in contrast to the statistics for the year before. The number of Turkish flagged vessels is likely to continue to decline in the near future.

Table 10: Recent Developments in the Turkish Oil Tanker Fleet

	Number of vessels	% Change	dwt (000)	% Change
1994	100	-	1,668	-
1995	99	-1	1,697	1.7
1996	103	4	1,609	-5.1
1997	98	-4.85	945	-44
1998	98	0	803	-15
1999	85	-13.2	1,418	76.5

Source: Republic of Turkey Prime Ministry Undersecretariat of Maritime Affairs, 2000.

Table 11: Oil Handling in Turkish Ports (000 tons)

From/To Turkish Ports	1998			1999		
	Turkish Flag	Foreign Flag	Total	Turkish Flag	Foreign Flag	Total
Loading	214,765	98,454	213,219	138,627	317,500	456,127
Discharging	9,075	11,279	20,357	8,890	9,454	18,343

Source: Republic of Turkey Prime Ministry Undersecretariat of Maritime Affairs, 2000. Unpublished statistics. (Cabotage carriages are not included).

The Turkish tanker fleet has faced a heavy loss for the last five years, and the number of oil tankers flying the Turkish flag has reduced significantly from 100 in 1994 to 85 at the end of 1999. However in 1999, the severe 1,465,000 dwt loss between 1995 and 1998, was replaced by an increase of 76% when compared with the year before and the Turkish oil tanker fleet has reached 1.4 million dwt (Table 10).

The problem with the Turkish tanker fleet, other than the shortage of tankers in number and tonnage, is that it is rather old. The average age is 25 years, which represents a serious problem in international trade stemming from new international legislation such as MARPOL, which commonly imposes a special survey at the age of 25 years.

Whilst the need for young tonnage is stronger than ever in order to play a role in international transportation, measures have to be taken to attract young tonnage to the Turkish fleet.

New Deliveries for Turkish Tanker Owners

The revival of Asian economies at the end of 1999 and promising developments for the Caspian Main Export Route and prospects for future market improvements raised the hopes of Turkish tanker owners.

In the mid and long term, following the anticipated elimination of older tankers from international trade, and a projected 45 million tons of crude oil flow to Turkey's Port of Ceyhan from the Caspian Region, there are indications of a strong crude oil transportation market for the Mediterranean.

The assumptions made concerning the crude oil market in the Mediterranean following construction of the Baku-Ceyhan Pipeline, encouraged tanker owners and others to invest in new crude oil tankers. Recent developments in the Turkish merchant fleet included three tanker owning and operating companies ordering crude oil tankers to be delivered late or after the year 2000. Two further crude oil tankers of 300,000 dwt each, are to be delivered by the end of 2000 and another newly established company has firm orders for ten bulk carriers in Japanese shipyards and is also considering new orders for oil tankers.

Conclusion

The current oversupply of 20% in the oil tanker market is expected to come to a balance in the short term. Although new deliveries in 2000 will constitute a higher supply/demand imbalance, this could change in favour of tanker owners following reduction in tonnage through scrapping and an increase in oil consumption from Asia and other countries (Table 4).

Meanwhile, oil prices are expected to increase through to 2015. This date is also likely to be when Caspian oil production will reach its highest level (Pala and Engur, 1998). This positive market trend will satisfy oil companies as they see the fruits of their investments with high profit margins. These investors have spent huge sums on exploration and production of the region's oil reserves and are enthusiastic to see the benefits. For example foreign investment in Azerbaijan's oil sector alone increased from US$15 million in 1993 to US$546 million in 1996, US$1.3 billion in 1997, and US$1.6 billion in 1998, equivalent to about 40% of Azerbaijan's GDP. Azerbaijan's Ministry of Economy has projected that foreign investment in the oil and gas sector could reach US$23 billion by 2010 (U.S. Energy Information Administration, 2000a).

It is obvious that no investor would put money forward for development of Caspian oil reserves unless the money spent shows a good return in due time. There is no doubt that no matter how high investments have been made for production, without suitable transportation alternatives to world markets, success cannot be achieved. Both pipeline and tanker shipments to and from the Port of Ceyhan will help the land-locked region to deliver its products to the world, and the multi-national and regional oil exploration companies and construction and transportation enterprises will all benefit.

The Baku-Ceyhan Pipeline Project has often been criticised due to the amount of oil projected to flow. Such a criticism may be considered right as the projected 45 million tons of oil may not be considered important when compared to total world demand. The construction of the pipeline is planned to be finished by 2004, and the amount of oil is projected to increase gradually from 11.5 million tons to 45 million tons in six years time. In 2010 world oil consumption is projected to increase by 32.3% from 3.4 billion tons in 1997 to 4.5 billion tons (Pala and Engur, 1999). Although the Baku-Ceyhan Pipeline will carry only 1% of world oil demand in 2010, the oil that will be brought to the East coast of the Mediterranean will positively affect the tanker market in the region.

Caspian oil is expected to be marketed mostly within Europe as its physical specifications are suited to the European Union's environmental standards. A market for 322 million tons of oil exists in Europe where Azerbaijan and Kazakhstan oil can compete and take its share (Pala and Engur, 1999).

It is not only the oil producers who will attempt to benefit from this new market, as also the tanker owners will also compete to play their role in the transportation of Caspian oil to world markets. The oil that will be shipped from the Turkish Port of Ceyhan and most likely delivered to Europe, will be transported largely by Turkish tankers.

The new optimistic era expected after the completion of the Baku-Ceyhan Pipeline, should encourage Turkish tanker owners to invest in new ships within the next four years, in order to achieve the desired potential in Turkish tanker shipments.

References

BP Amoco (2000) *Statistical Review of World Energy, 1999 in Review*, Chicago: BP Amoco Publication.
Cohen, A. (1997) US Policy in the Caucasus and Central Asia: Building a new "Silk Road" to Economic Prosperity, *Backgrounder*, No:1132, July 24.

Crow, P. (1998) Caspian Dreams. *Oil and Gas Journal*, October 26, 96, 43.

Crow, P. (1999) Watching Government Caspian Options. *Oil and Gas Journal*, November 1, 97, 44.

Dion, R.R. (1999) Long view of Caspian Oil Export Options Tilts to Kazakhstan-China, *Oil and Gas Journal*, 7, 97, 23.

Doll, F. (1999) Tanker Market Overview, *BIMCO Review 1999*, London: BIMCO Publication.

Drewry (1992) *Oil Trades, Secondary Transportation Modes and Tanker Demand*, London: Drewry Publications.

Drewry (1993) *The Aframax Tanker Market. A Strategic and Commercial Assessment to 2005*, London: Drewry Publications.

Fields, C. (2000) Demolition and Scrapping Slow after Busy Week, *Lloyd's List*, April 28.

Fields, C. (2000) Non-Opec Nation Feel Pressure, *Lloyd's List*, March 29.

Gezgin, E. (1999) *Caspian Oil and Transportation to the Market*, University of Plymouth, Institute of Marine Studies, Unpublished MSc Dissertation.

Grace, J.D. (1998) Caspian Production, Export, Investment Outlooks Sized up, *Oil and Gas Journal*, August 24, 96, 34.

Gray, T. (2000) Tankers and Post-Erika Scrapping Pressures Likely to Bring Sustained VLCC Rates Improvement, *Lloyd's List*, April 28.

Gülen, G. and Foss, M. (1999) Caspian Oil Export Choices Clouded By Geopolitics, Project Economics, *Oil and Gas Journal*, April 19, 97, 16.

Helmer, J. (2000) Novorossisyk Plans Capital Drive, *Lloyd's List*, April 24.

International Energy Agency (1998) *World Energy Outlook 1998*, Paris: IEA/OECD Publication.

International Energy Association (2000) *Monthly Oil Market Report*, June, Paris: IEA/OECD Publication.

Lowry, N. (2000) Brussels Bid on Oil Spill Liability, *Lloyd's List*, June 5.

McMahon, A. (2000) Review of the Tanker Market 1999. *BIMCO Review 1999*, London: BIMCO Publication.

McMichael, B. (2000) Lukoil Joy in North Caspian Waters, *Lloyd's List*, March 25.

Pala, C. and Engür, E. (1998) Kafkasya Petrolleri 21. Yüzyilin Esiginde Hazar Havzasi ve Türkiye, *Isletme ve Finans*, Sayi 152, Kasim.

Pala, C. and E. Engür, (1999) Yüzyil Dünya Enerji Dengesinde Petrolün ve Hazar Petrollerinin Yeri ve Önemi, *PetroGas Dergisi,* Sayi:11, Mart-Nisan.

Peters, H. (2000) The World Economy and Sea Trade. *BIMCO Review 1999,* London: BIMCO Publication.

Republic of Turkey Prime Ministry Undersecretariat of Maritime Affairs, 2000, Unpublished statistics.

Salles, B.R. (1999) Shipping and Shipbuilding in 1998, *BIMCO Review 1999,* London: BIMCO Publication.

SPO - State Planning Organisation, 2000. *8th Five-Year Development Plan 2000-2005 Report Drawn up by Special Ad Hoc Committee on Maritime Transport,* Ankara: Unpublished Report.

TCS-Turkish Chamber of Shipping (1999) *Deniz Sektoru Raporu 1998,* Istanbul: Chamber of Shipping Publication.

Tutt, N. (2000) Brussels Tanker Rules Expected by Year-end, *Lloyd's List,* April 20.

UNCTAD (1999) *Review of Maritime Transport 1999,* Geneva: UNCTAD Publication.

US Energy Information Administration (1999) *Turkmenistan,* Washington DC: Unpublished report.

US Energy Information Administration (2000a) *Caspian Sea Region,* Washington DC: Unpublished report.

US Energy Information Administration (2000b) *International Energy Outlook 2000,* Washington DC: US Energy Information Administration Publication.

US General Accounting Office (1994) International Trade, Kazakhstan Unlikely to be Major Source of Oil for the United States, *Report to Congressional Requesters.* Washington DC: Unpublished report.

World Oil (1999) *54th Annual International Outlook,* August.

Young, T. (1999) Recent Market Trends and Their Impact, *BIMCO Review 1999,* London: BIMCO Publication.

9 Turkish Shipbuilding in the 1990s

ERHAN BAYRAKTAR, HARRY HEIJVELD AND MICHAEL ROE
INSTITUTE OF MARINE STUDIES
UNIVERSITY OF PLYMOUTH

Introduction

Shipbuilding has long been a growing and developing industry in Turkey and along with the shipping industry, has played a substantial role in the national economy. The shipbuilding sector is also one of the major sectors of international activity within the Turkish economy. It is traditionally included within the manufacturing industry subsector of the Turkish Five-Year Development Plan.

The shipbuilding industry involves manufacturing and assembly, provides a considerable source of foreign exchange, encourages a number of auxiliary industries, attracts the transfer of technology, provides for a range of employment opportunities, supports the national commercial fleet and contributes significantly to the needs of national defence (Nehir, 1993).

The shipbuilding industry earns foreign exchange as long as it continues to export its products. The industry reduces the import needs for newly built vessels from other countries and therefore reduces the loss of foreign exchange. The development of an industry that can contribute directly to foreign exchange acquisition and conservation within the national economy is considered in Turkey with some importance. Therefore, the shipbuilding industry plays an important role in Turkey, a country that commonly exhibits economic problems through severely imbalanced foreign trade.

The design of ships varies according to the seas in which they are expected to operate, with their purpose, their principles of operation, in the way they are propelled, with the material used in their construction, their tonnage and the technical facilities they have on-board.

A ship is essentially a floating vehicle, built in shipyards for the purpose of carrying people and/or cargoes to achieve some particular service on water. It is constructed by bringing together the necessary components and equipment to the shipyard, by producing various components within the shipyard to be placed in the body of a ship, all in accordance with the planned project.

These components include those of various sectors, such as iron and steel, machinery, paints, products of wood, plastics and electronics. A ship is the product of the shipbuilding industry and is used in shipping and transport operations. Therefore, the shipbuilding industry forms a major substructure of sea transportation. The shipbuilding industry of Turkey consists of the construction, restoration, alteration and maintenance of all types of floating vehicles.

A Brief History of the Turkish Shipbuilding Industry

The history of the Turkish shipbuilding industry is outlined below in two sections - shipbuilding during the Ottoman Empire era, between the years 1214-1923; and shipbuilding after the establishment of the Republic of Turkey in 1923.

Shipbuilding in the Ottoman Empire Era

In the Seljuk era, the shipyards of Sinop and Alanya, established in 1214 and 1227 respectively, were very significant. In addition, during this era, amongst the shipyards on the Anatolia peninsula, the Izmir shipyard was notable as an important one. The Izmir shipyard was established in 1326 and belonged to the Principality of Aydin. Many of these large shipbuilding facilities had developed parallel to the enlargement of the Ottoman Empire. In the establishment phase of the Ottoman Empire, Karamursel, Gemlik, Aydincik and Gelibolu shipyards should also be mentioned as the more important ones. Additionally, Izmit was of some note and Gelibolu shipyard had an annual capacity of 15 galleys, the ubiquitous Ottoman style, wooden vessels.

During the developing era of the Ottoman Empire, Istanbul, Suez and Ruscuk shipyards had been added to the shipyard industry of the time. The more important were the Istanbul shipyards that were established by Sultan Mehmet the Conqueror in 1455. These shipyards continue to operate at present under the names, "Halic", "Camialti" and "Taskizak" and belong to the Turkiye Gemi Sanayi A.S. (Turkish Ship Industry Company). These shipyards had the capability of building the largest ships possible during their early days (Chamber of Naval Architects, 1989).

From the 16th century to the collapse of the Ottoman Empire in 1920, there was a generally increasing economic crisis throughout the Empire. Thus many new developments in the world economy missed Turkey during this period and this also influenced the development of the shipbuilding industry. In spite of this, the shipyards were able to keep the Ottoman fleet active without any foreign support. The most important technical developments in the Turkish economy commonly emerged from the shipyards.

Meanwhile, the Emperor School of Naval Engineering was established in 1773 near the Istanbul shipyards, by specialists from a number of European countries. This institution acted as the foundation for Istanbul Technical University, the University of Yildiz and the Military Maritime Academy which still exist today (Chamber of Naval Architects, 1995).

In the nineteenth century, with the continued industrialisation of western economies, steam engines began to be installed on ships. The Ottomans had taken precautions to adapt to this trend by obtaining the appropriate technical knowledge and training suitable manpower in order to keep up with this development. The stone dry-docks in Halic shipyard and the Valide slide in Taskizak were constructed in the first half of this century whilst other facilities such as rolling mills, foundries, ironworks, paint shops and steel furnaces were also completed.

However, the shipyards began to lose their ability to support the fleet in spite of these developments. The main reason for this was the dependence upon foreign supplies in the steel ship era. Additionally, the domestic industry in general was unable to give further support to the shipping industry and whereas both raw and semi-manufactured materials were formerly supplied from domestic resources, this now was increasingly less the case.

The Republic from 1923

When the Republican age began in 1923, the Halic and Taskizak shipyards had ceased their military activities following the articles of the Lausanne Agreement related to the two straits, the Bosphorus and Dardanelles. The

Haskoy shipyard was inactive. The most significant problems of the time lay in front of the government and these included developing shipyard facilities to meet the needs of the national economy, raising the Naval fleet to a higher efficiency level, and developing merchant marine transportation. For the first time in 1924, a floating dock was constructed by a German company at a Turkish shipyard and the same company constructed a shipyard in Golcuk. During the same period, Taskizak shipyard and some equipment of the Halic shipyard was moved to Golcuk shipyard. A state owned institution called "Turkish Shipyard and Dock; Workshops and Naval Factories Directorate" was established in 1928. Some projects were prepared by a Dutch company in order to convert this establishment into a modern shipyard, but this project stopped at the commencement of World War II. It was restarted in 1942.

In 1933, the Halic Shipyard was converted into the Factory and Dock Operations Directorate and continued its development. In this period, the yard started to build ships with small dimensions. Two passenger-cargo ships for Van Lake operations in the eastern region of Turkey, various tug boats, pilot boats, service boats and a pier for the Kadikoy-Haydarpasa line were constructed during this period.

In 1933, operation of the shipyards was taken over by the State Maritime Lines and Harbours General Directorate, whilst in 1936, Taskizak shipyard started operations again following the Montreux Agreement. This shipyard was converted into one for the Navy and submarines for German clients were constructed.

In 1939, it was considered that the existing shipyard development possibilities were limited and it was decided to construct a shipyard at Pendik Bay, which could build 50,000 dwt a year in order to supply the needs of the country. For this purpose, students were sent to England and Germany to train in technical subjects. However, these plans could not continue due to the outbreak of World War II but those students that had passed through this programme, became the core of the future shipyard operations. Meanwhile, a Shipbuilding Faculty was established at the Istanbul Technical University, and Camialti, Haskoy and Taskizak shipyards started to operate in 1944.

Between 1950-1963, the shipbuilding industry, which had been limited to building small vessels, barges, buoys, various mooring boats and tug boats, started to build larger vessels. Examples included 6,500 dwt general cargo ships, ferries, various tug boats and passenger ships. In 1962, a floating dock with 10,000-15,000 ton lifting capacity was constructed

(Chamber of Naval Architects, 1989).

The period after 1963 is known as the Planned Period because of the Five-year Development Plans that were introduced in Turkey. Progress that has been made in the Turkish shipbuilding industry during this period is explained in the section upon recent developments within the Turkish shipbuilding industry to be found later in this paper.

Shipbuilding in the World and Turkey

Almost 95% of world trade is transported by sea and the demand for sea transport keeps increasing parallel to the development of world trade. Ships, which are the instruments of sea transport, have limited useful lives particularly regarding technical and economic aspects. Under heavy working conditions and with the corrosive effects of sea water, ships wear away and their maintenance and repair costs increase as they age and therefore they lose their economic productivity. In particular, shipowners dispose of their vessels and renew their fleet when freight rates are profitable and this of course leads to renewed demands for shipbuilding.

Moreover, as a result of the extended use of intermodal transport and changes in world trade, new types of vessels have started to be built using high technology. Increasing productivity of the newly designed vessels has increased the demand for newbuildings to take advantage of the efficiency gains that can be realised.

Figure 1: World New Vessel Building Orders, 1983-1993

Year	Orders (million gross tons)
1983	32.6
1984	30.7
1985	25.9
1986	21.4
1987	22.5
1988	24.6
1989	31.1
1990	29.8
1991	43.2
1992	37.3
1993	39.2

Source: Drewry Shipping Consultants, 1995.

Newbuildings of ships world-wide are forecast to be approximately 300 million dwt up to the year 2001 and the new investment approximately US$ 350 billion (State Planning Organisation, 1995).

Demand for newbuildings however, fluctuates due to the politics of pricing and the variability of the freight market. Variations of orders for new vessel building in the world are illustrated in Figure 1.

Figure 2: World Shipbuilding Output and New Orders, 1983-1993 (million gross tons)

Year	Output (million grt)	Orderbook (million grt)
1983	15.6	32.6
1984	18.3	30.7
1985	18.2	25.9
1986	16.8	21.4
1987	12.3	22.5
1988	10.9	24.6
1989	13.2	31.1
1990	15.9	29.8
1991	16.1	43.2
1992	18.6	37.3
1993	19.8	39.2

Source : Drewry Shipping Consultants, 1995.

Supply and Demand for Shipbuilding in the World

Over the past three decades, shipbuilding has progressed from an unsophisticated industry to one of high technology. Meanwhile, the world shipbuilding industry has become plagued by overcapacity and therefore, the industry has had to undertake various steps to bring production closer to demand. The majority of the surviving shipyards in the world are those that can still depend upon state subsidy. Many problems arise on the supply side of newbuildings due to incorrect market predictions that lead to oversupply in the market and an excess of tonnage. Demand for different types of ships has also been influenced by structural and regional changes in world trade. The forecasting of changes in demand has become more complex because of the influence of many factors which continuously shift in their relative importance. Furthermore, since shipping supply is intensely inelastic in the short term, it is very difficult to get supply and demand in balance (Drewry, 1995).

In Turkey, the growth of the fleet that has characterised recent years has not been the result of new investments and the effect of this has been a notable increase in age profile.

The level of newbuilding can be measured either in terms of completed tonnage or in terms of the size of new orders. Figure 2 illustrates annual world shipbuilding output in gross tonnage, and the total tonnage of orders, between 1983 and 1993. As can be seen, the world shipyard output was at its lowest level between 1986-1988, which covered a severely weak freight period in the world. However, as ordering activities for newbuilding increased during the late 1980s, the output recovered to 15.9 million grt in 1990. Furthermore, this amount continued to increase and reached 19.8 million grt in 1990. New orders have also increased during this period between 1986-1989 and reached 39.2 million grt in 1993.

Shipyard utilisation rates also increased in the 1990s. Additionally, shipbuilding demands started to require physically bigger shipyards. Therefore, expansion plans were made for certain shipyards in the world, including major developments in Korea (Drewry, 1995).

1994 was a dramatic year for the shipbuilding industry. Japan, as the world leader, reacted against Korea, which was increasing capacities of their shipbuilding and shipyards. Additionally, OECD countries decided to cut the loans for the building industry from January 1996. This severely affected the shipbuilding industry in the world and significant fluctuations were seen after 1994.

On the other hand, some European shipbuilders gained greater profit due to changes in the valuation of the Japanese Yen against the US$. However, they had limited success in the building of standard tankers and bulk cargo ships and therefore, moved from these products to a more specialised form of shipbuilding.

Scandinavian countries proved more powerful in building chemical tankers and have become highly successful in the shipbuilding industry in the world as a result. Polish shipyards have placed more importance upon the building of container ships. The differences between European shipbuilders have become more severe recently. For example, whilst some Danish shipbuilders face an uncertain future in the market others have become very successful.

Turkish Shipbuilding Industry in the World

The crisis in the world sea transport market during the mid 1980s caused vessel prices generally to decrease. The Turkish shipbuilding industry was also affected by this crisis. In particular, imbalances within the Turkish national economy led to a decrease in subsidies for shipowners and for the shipbuilding industry.

Figure 3: Shipbuilding Output and Orders in the World (01.01.1995)

Country	Under Construction		On Order		Total	
	Pieces	grt x 1,000	Pieces	grt x 1,000	Pieces	grt x 1,000
EU	142	3,103.8	191	4,330.1	333	7,433.9
Germany	43	736.0	52	996.2	95	1,732.2
Italy	21	483.2	19	702.5	40	1,185.7
Finland	9	547.7	9	501.5	18	1,049.2
France	8	472.5	3	257.0	11	729.5
Denmark	5	213.4	32	828.2	37	1,041.6
Spain	12	272.5	26	457.5	38	730.0
Netherlands	24	153.0	28	107.2	52	260.2
UK	8	139.2	10	331.7	18	470.9
Greece	5	66.4	1	5.0	6	71.4
Belgium	2	10.5	2	111.6	4	122.1
Portugal	5	9.5	8	31.1	13	40.6
Sweden	0	0	1	0.7	1	0.7
Other OECD	234	5,698	293	9,919	527	15,618
Japan	193	5,370.2	252	9,464.9	445	14,825
Norway	11	135.7	17	184.5	28	320.2
Turkey	19	169.6	15	208.0	34	377.6
USA	1	0.7	7	57.6	8	58.3
Australia	9	22.1	2	4.4	11	26.5
NZ	1	0.4	0	0	1	0.4
Total OECD	376	8,802	484	14,249	860	230,551
Total orders %	43.7	38.2	56.3	61.8	100.0	100.0

Source: ISL, 1995.

Output from Turkish shipyards rose to 68,000 grt in the late 1980s, representing 2% of the total for West Europe. This was largely achieved with the encouragement of state funds which had restarted after the earlier economic crisis and a good number of orders largely from German and Polish shipowners. Shipbuilding output and new orders from the EU and other OECD countries at the beginning of 1995, are indicated in Figures 4 and 5 which show that total Turkish shipbuilding output, together with the new orders at the beginning of 1995 was 34 vessels and 377,600 grt. Turkey was the seventh country in the world with 3.95% of newbuildings and new orders.

Figure 4: World Shipbuilding Output and New Orders by Vessel Type (pieces)

Country	Tanker	Dry bulk	Container	General cargo	Cruiser	TOTAL
EU	68	27	64	91	83	**333**
Germany	7	2	53	22	11	**95**
Italy	14	7	0	1	18	**40**
Finland	4	1	0	0	13	**18**
France	4	0	0	0	7	**11**
Denmark	12	8	3	9	5	**37**
Spain	10	2	4	13	9	**38**
Netherlands	4	1	4	39	4	**52**
UK	4	4	0	2	8	**18**
Greece	0	0	0	1	5	**6**
Belgium	4	0	0	0	0	**4**
Portugal	5	2	0	4	2	**13**
Sweden	0	0	0	0	1	**1**
Other OECD	124	188	61	105	49	**527**
Japan	106	186	55	83	15	**445**
Norway	11	0	1	3	13	**28**
Turkey	5	2	4	17	6	**34**
USA	2	0	0	1	5	**8**
Australia	0	0	0	1	10	**11**
NZ	0	0	1	0	0	**1**
Total OECD	**192**	**215**	**125**	**196**	**132**	**860**
Total Order %	**22.3**	**25**	**14.5**	**22.8**	**15.3**	**100**

Source: Government of Turkey, 1995.

As illustrated in Figures 4 and 5, Turkish shipbuilding output and new orders compared with world shipbuilding output and new orders were as follows during this period:

- 5 tankers out of a total of 192;
- 2 bulkers out of a total of 2,152;
- 4 container vessels out of a total of 125;
- 17 general cargo vessels out of a total of 196;

- 6 passenger ships/ferries out of a total of 132.

Figure 5: World Shipbuilding Output and New Orders by Vessel Type (000 grt)

Country	Tanker	Dry bulk	Container	General cargo	Cruiser	TOTAL
EU	2,741	966	1,189	431	2,107	7,434
Germany	72	40	1,094	86	440	1,732
Italy	451	184	0	8	543	1,186
Finland	461	10	0	0	578	1,049
France	401	0	0	0	328	729
Denmark	644	296	29	52	20	1,041
Spain	354	162	32	115	67	730
Netherlands	12	15	33	146	55	261
UK	204	251	0	3	13	471
Greece	0	0	0	9	62	71
Belgium	122	0	0	0	0	122
Portugal	20	8	0	12	1	41
Sweden	0	0	0	0	1	1
Other OECD	6,000	6,219	2,304	843	251	15,617
Japan	5,611	6,135	2,281	701	107	14,835
Norway	173	0	8	33	106	320
Turkey	164	84	15	106	9	378
USA	52	0	0	1	6	59
Australia	0	0	0	3	24	27
NZ	0	0	0	0	0	0
Total OECD	8,742	7,185	3,493	1,274	2,358	23,052
Total Order %	37.9	31.2	15.2	5.5	10.2	100.0

Source: Government of Turkey, 1995.

Furthermore, Turkish shipbuilding output and new orders compared to world figures in gross tonnages were as follows:

- 164,000 grt of tankers from 8,742,000 grt;
- 84,000 grt of bulkers from 7,185,000 grt;
- 15,000 grt of container vessels from 3,493,000 grt;
- 106,000 grt of general cargo vessels from 1,274,000 grt;
- 9,000 grt of passenger ships/ferries from 2,358,000 grt.

There remained an urgent need to upgrade the shipbuilding industry in Turkey because of ageing facilities and the relatively poor technology of the shipyards compared with those in developed countries. As a result, steps needed to be taken to improve productivity and to make arrangements to improve the financial situation. Furthermore, a consortium of Turkish shipowners made plans to import Japanese shipbuilding technology into local shipyards together with a loan from Japan (Drewry, 1995).

Figure 6: Turkish Merchant Fleet by Type (% dwt)

Vessel type	1994	1998
Bulk carrier	50.0	65.8
General cargo	17.1	13.6
Tanker	19.4	9.1
OBO	10.6	6.3
Ro-ro/ferry	0.4	1.6
Container	0.4	1.1
Others	2.1	2.5
Total	100.0	100.0

Source: Chamber of Shipping, 1995; Forum, 1998.

The Relationship Between the Turkish Shipbuilding and Shipping Industries

Turkey, as a peninsula with approximately 8,000 kilometres of sea coast, has enormous potential to compete in shipping markets with other countries in the world. The total tonnage of the Turkish merchant fleet was 5,049,571 grt totalling 1,141 vessels at the end of 1993. The fleet dropped to 1,050 vessels by the end of 1994.

The percentages of types of vessels in the Turkish merchant fleet in 1994 and 1998 are shown in Figure 6. The general development of the fleet from 1985 to 1994 is shown in Figure 7. The average ages of various vessel types in the Turkish merchant fleet are illustrated in Figure 8.

As can be seen from Figure 8, the Turkish merchant fleet mostly consisted of vessels over the age of 15 years. The shipbuilding industry in Turkey plays an important role due to the ageing of the merchant fleet and the pressing need for its renewal combined with an increasing demand for repairs as the age profile deteriorates (Tez, 1993).

Figure 7: Development of the Turkish Merchant Fleet 1985-1994 (000 dwt)

Vessel	1985	1986	1987	1988	1989	1990	1991	1992	1993	1994
Dry cargo	1,271	1,309	1,393	1,388	1,340	1,335	1,367	1,344	1,404	1,468
Bulk	1,783	1,922	1,838	1,740	1,790	2,266	2,396	2,719	3,842	4,273
OBO	358	519	519	286	384	384	534	534	792	961
Crude oil tanker	2,285	1,374	1,377	1,380	1,531	1,559	1,560	1,778	2,077	1,667
Chemical tanker	50	46	55	58	57	36	38	47	60	78
LPG	7	7	7	7	7	7	11	11	13	13
Asphalt tanker	5	4	4	4	4	6	6	8	8	8
Fresh water tanker	2	3	3	3	3	3	4	4	4	6
Ro-ro	15	15	15	16	16	14	21	27	24	79
Container	0	0	4	4	4	0	0	0	0	5
Ferry	8	8	10	8	8	8	10	9	9	10
Train ferry	6	6	7	7	7	7	7	7	7	7
Livestock carrier	0	4	0	0	2	0	0	0	0	0
Reefers	6	6	4	0	0	0	0	0	0	0
Others	13	9	5	8	12	14	14	14	13	14
Total	5,808	5,234	5,240	4,911	5,166	5,639	5,968	6,503	8,255	8,590

Source: Government of Turkey, 1995.

Figure 8: Average Ages of Vessels in the Turkish Merchant Fleet (end 1994)

Type	State Imports	State Newbuildings	Private Imports	Private Newbuildings	Overall Average
Dry cargo	24.4	11.6	25.9	12.2	**15.5**
Bulk	20.1	13.6	17.8	0	**17.9**
OBO	0	0	20.7	0	**20.7**
Oil tankers	19.9	17.7	25.2	14.6	**21.0**
Chemical tankers	0	0	22.2	13.5	**19.9**
LPG	0	0	14.1	22.6	**15.8**
Asphalt tankers	0	33.6	31.4	0	**32.5**
Ro-ro	10.4	0	16.5	0	**14.1**
Container vessels	0	0	0	0.5	**0.5**
Ferries	10.6	3.4	0	0	**8.2**
Small passenger boats	39.6	0	0	0	**39.6**
Overall Average	11.4	7.3	15.8	5.7	**18.7**

Source: Government of Turkey, 1995.

Structure of the Turkish Shipbuilding Industry

Recent Developments in Turkish Shipbuilding

During the planned economy period in Turkey, which commenced from 1963, development of the maritime industry ran parallel with the development of the economy. During the First Five-Year Development Plan between 1963-1967, the main target for the shipbuilding industry was to provide the ships that were needed by Turkish operators, to be constructed by domestic shipyards. The number of imported ships could thus be minimised during the early years of this period between 1963-1965. A variety of vehicle ferries, passenger ferries and some merchant ships were constructed in Turkish shipyards as a result.

After 1965, ship demand filled shipyards in Turkey and full capacity was achieved. As a result, a number of public sector shipyards were developed and some private sector shipyards newly established. A number of credit mechanisms were introduced during this time and substantial amounts of credit were supplied to the industry. During the same period, much needed investment was carried out in order to bring the Camialti shipyard up to a standard to build modern ships. This shipyard was provided with a slide that enabled ships up to 20,000 dwt to be serviced and a series of modern workshops was completed in 1970.

During the Second Five-Year Development Plan between 1968-1972, steps were taken to ensure that the shipbuilding capacity in the country was increased and modernised with the basic aim to supply the demand for ships from domestic resources. However, to provide for rapid growth and to avoid congestion within the shipbuilding industry, dry cargo vessels, large tankers and other specialist types of ships could be imported. Even so, domestic ship demands could not be met by the existing shipyards both because of increasing ship dimensions and annual over-demand for production.

This situation had been exacerbated by continued delays in investment - the Pendik shipyard for example, planned to be built in 1939, was finally included in the 1969 investment plan - and in 1974, the banks began to provide credit for investment and support for foreign supplies. However, finance for investment in the dry dock was of such a size that it was decided to build only half of it along with the necessary slide, facilitating the construction of ships only up to 60,000 dwt. The slide was completed in 1982.

Amongst the targets included in the Third Five-Year Development Plan between 1973-1977, was setting up the necessary systems and organisations so that the Turkish merchant marine fleet could service foreign markets and owners. The purpose of having an effective merchant fleet was to reduce the payment of high cargo prices through dependence upon foreign owners and to gain a foothold in foreign markets and the lucrative hard currency trades that existed. It was also a target for the Turkish shipbuilding industry to supply the entire domestic demand and to be developed to supply the export market. With these considerations in mind, the Alaybey Yard was also to be developed as a ship repair facility for the Aegean region and was thus included in the investment programme.

It was planned for 1975 that the Turkish merchant marine fleet and shipbuilding industry would supply 50% of Turkish foreign trade and where necessary, vessel needs which could not be supplied from domestic production would be imported (Chamber of Naval Architects, 1989).

During the same period, the capacity of the shipyards was increased and as a result, demand could be accommodated from domestic resources apart from a number of very large tankers and a limited number of specialist vessels.

During the period of the Fourth Five-Year Development Plan between 1979-1983, the shipbuilding facilities in Tuzla in the Marmara region, were developed as an alternative centre to the shipyards of Istanbul. As a result of these improvements between 1981-1983, 89 ships totalling 220,000 dwt and ranging between 1,500 and 75,000 dwt were constructed at shipyards in this region for operators from the private sector.

However, following this period, shipbuilding in private sector shipyards stopped completely and the industry generally fell into crisis from 1985. This continued until the end of 1987 when in early 1988, the Central Bank of Turkey postponed the credit debts of the Marine Bank and the latter started to make repayment agreements with the shipbuilders. This cleared the financial obstructions to progress and as a result, in the next few years, one of the private sector shipyards had orders for 24, 4,200 dwt ships, and another a contract to build a series of steel barges for a firm from the Netherlands.

Other shipyards negotiated for orders from overseas. Turkish shipbuilding prices were approximately 20% cheaper than other countries in spite of the technological inefficiencies that remained and the tendency towards longer construction times.

With the separation of state shipyards from the Marine Bank and the establishment of Turkiye Gemi Sanayi A.S Genel Mudurlugu (General Directorate of Turkish Ship Industry Company), the state owned shipyards started to be operated by the Turkish Ship Industry Company. At this stage the largest ships that were able to be built in the country were the two 75,000 dwt bulkers for the Turkish Marine Transportation Company constructed at Pendik shipyard. The engines (Pendik-Sulzer [A Type]) were produced under license from Sulzer and could generate up to 35,000 BHP.

Regarding the report on the shipbuilding industry of the Sixth Five-Year Development Plan for the Turkish economy (1989-1993), the target for total tonnage was determined by predicted imports and exports of the country. The total tonnage of the Turkish merchant fleet was targeted to reach 6.5 million dwt including approximately 1.5 million dwt of newbuildings (Chamber of Shipping, 1995). However, actual domestic newbuildings were much less than the planned quantities and this differential is illustrated in Figure 9.

Figure 9: Planned and Actual Shipbuilding in Turkish Shipyards, 1989-1993 (000 dwt)

Year	Planned	Actual
1989	100	10
1990	100	10
1991	305	30
1992	350	30
1993	400	5

Source: State Planning Organisation, 1995.

Current Situation in Turkish Shipbuilding

The current (1999) total potential capacity of the Turkish shipyards is 73,000 dwt/year for new buildings and 6 million dwt/year for ship repairs. However, the actual usage of capacity is between 10 and 20%, which is considered too low.

Since there are strong links and a high dependence between the primary and secondary shipbuilding industries, total employment for the sector is not limited to the main shipbuilding industry. It is estimated that one person employed in the primary shipbuilding industry provides employment for three people in the secondary shipbuilding industry in Turkey (Chamber of Shipping, 1995a).

The Technical Situation

Wooden, steel, aluminium and fibreglass vessels have been constructed in Turkish shipyards. The largest number of enterprises that build steel-aluminium, wooden and fibreglass ships are now gathered in Istanbul and the Marmara region but there remain some units in Karadeniz Ereglisi, Izmir and Tatvan. Generally, wooden shipbuilding has been concentrated on the Black Sea coast (Trabzon and Zonguldak), in Marmara (Tuzla, Sariyer, Kucukcekmece and Selimpasa), on the Aegean coast (Izmir and Fethiye), and on the Mediterranean coast (Bodrum, Marmaris and Alanya). Fibreglass vessel production is carried out in Istanbul and Izmir. Wooden and fibreglass construction is used for small vessels and yachts, mainly for leisure purposes that play an important role for the Turkish tourism sector but which are of much less consequence in the Turkish economy.

Therefore, only the technical situation of shipyards concentrating upon steel and aluminium vessels is discussed below.

Shipyards

The major Turkish shipyards are categorised by the Ministry of State in the following way:

- shipyards that belong to the Navy;
- state owned shipyards that belong to the Turkish Ship Industry Company (Turkiye Gemi Sanayi A.S.);
- shipyards that belong to the private sector.

Shipyards that belong to the Navy: The Navy is quite different from the other organisations involved in shipbuilding due to its broader defence duties. General functions are:

- keeping the ships of the naval fleet in an active condition;
- within budget possibilities, building war and auxiliary ships and equipment;
- to supply necessary technical services for shore facilities;
- to design and produce spare parts and machinery for ships and shore facilities;
- to supply from excess capacity, private sector shipbuilding, repairing, docking and technical services.

The *Navy* has two notable shipyards, one of which is Golcuk shipyard which at present is the most important unit in Turkey and the East Mediterranean. In this yard, submarines, destroyers, tankers and bulk material ships of up to 25,000 dwt can be built, repaired and overhauled by well qualified and skilled workers, using the widest range of specialised and large machine tools.

Shipyards that belong to the Turkish Ship Industry Company: There are five shipyards that belong to this state owned company. These are Halic, Camialti, Istinye and Pendik shipyards in Istanbul and Alaybey shipyard in Izmir. Additionally, there is an engine factory that produces diesel engines with a Sulzer license in Pendik shipyard.

Private sector shipyards: The number of private shipyards has increased rapidly with the subsidy and support available during the Fourth Five-year Development Plan, and particularly in Tuzla in the Marmara region, situated on the northwest coast of Turkey. However, because of the bottleneck in shipbuilding credits, imbalances and instability within the Turkish economy increased rapidly after 1985 creating unemployment in private sector shipyards. As a result many of the Tuzla shipyards could not complete their investment programmes and have not developed the required capacity.

Turkish shipyards thus need to develop technologically and open up to more foreign markets. Some examples of where this is happening include Marmara Transport Shipyard constructing river-sea tankers for Russia, Sedef Shipyard building various types of vessels for Russia and river-sea tankers for Azerbaijan, cargo vessels and chemical carriers for Germany and Norway, and Madenci Shipyard with a series of vessels for Denmark (Lloyd's List, 1993; Lloyd's Ship Manager, 1994).

When the machinery and equipment used in the shipbuilding and maintenance industries are considered, the related component industry is a vital part of the sector and is an important addition to the economy of the country. As a result, the level of subsidies must be reviewed in the light of their wider impact and the demand for domestic new shipbuilding and exports must be remembered. In the light of this, the annual new shipbuilding demand in 1995-2000 has been estimated as double the 1994 annual average, increasing total tonnage from 10 million dwt to 20 million dwt.

Figure 10 illustrates the specifications and capacities of private shipyards in Turkey. Specifications and capacities of the state owned shipyards that belong to the Turkish Ship Industry Company (part of the Turkish Shipbuilding Industry Inc.) are illustrated in Figure 11.

Regarding the capacities of Turkish shipyards belonging to both private sector and the state, capacity utilisation is approximately 10-20%, leaving approximately 80% over-capacity.

The Economic Situation

The impact of the current total capacity of the shipbuilding industry upon the Turkish economy can be summarised as follows (Chamber of Shipping, 1995a; 1995b):

- added value of the current capacity: US$ 1.2 billion per year;
- added value of the current capacity in terms of net exchange rate inputs: US$ 600 million per year;
- added value of the current capacity to employment (including the secondary shipbuilding industry): 105 million man-hours per year;
- revenue to the Turkish economy with regards to the usage of current total capacity (including taxes, insurance, funds, interest rates, etc.): US$ 340 million per year;
- rate of domestic added value: 40%.

It is widely agreed that if the current technical capacity of Turkish shipyards is increased, then the economic situation of the industry will also be affected. As a consequence, the following have been planned by the State Planning Organisation for the Turkish shipbuilding industry in the Seventh Five-Year Development Plan for the period between 1995-1999 (Chamber of Shipping, 1995a):

- new shipbuilding capacity: 1,000,000 dwt/year;
- ship repair capacity: 7,500,000 dwt/year;
- added value of the shipbuilding industry to the economy: US$ 1.5 billion per year;
- added value of total capacity to employment (including the secondary shipbuilding industry): 125 million man-hours per year;
- rate of domestic added value: 65%.

The following are recommended within the Seventh Five-Year Development Plan to achieve the above objectives during the period between 1995-1999:

- extended support for maritime education and training in shipyards;
- new shipbuilding projects and marketing of these projects;
- improvements to the substructure of shipyards;
- new subsidies;
- privatisation of state owned shipyards;
- establishment of a specialised sector bank.

The Financial Situation

Regarding the Seventh Five-Year Development Plan between 1995-1999, the total capacity of the Turkish merchant fleet was planned to reach approximately 20 million dwt from 10.3 million dwt at the end of 1995.

This total was estimated to reach 15 million dwt by the middle of the planned period (Chamber of Shipping, 1995b). Furthermore, this meant that an average of 1.5 million dwt of vessels was planned to join the Turkish merchant fleet each year during this period. If 10-15% of the new vessels in the fleet were assumed to be newbuildings, then approximately 150,000 dwt of newbuildings were planned for each year during this period (Chamber of Shipping, 1995d).

Additionally, the required finance for the building of a new ship was calculated as around US$ 1,000 per dwt (UNCTAD, 1994). Therefore, the total annual financial requirement for newbuildings by the Turkish shipbuilding industry during the planned period between 1995-1999 became US$ 150 million (Chamber of Shipping, 1995d). The annual financial requirement from the scrapping of old ships was estimated as US$ 100 million. In addition to this, there was another financial requirement, which was for second-hand ship acquisition. The annual financial requirement for this purpose was thus estimated as a minimum of US$ 250 million. Therefore, the Turkish shipbuilding industry needed a total of US$ 500 million each year during the period of the Seventh Five-Year Development Plan, to achieve its targets for the industry.

Figure 10: Comparison of Subsidies for the Shipbuilding Industry in the World and Turkey

Subsidy	World	Turkey
Official	9-26%	10% + VAT
Hidden	Available	Not available
Long term loans	Available	Not available
Exchange rate - actual rate difference	Not available	Available
Warranty	Based on project	Property, land, vessel
Specialised banks	Available	Not available

Source Chamber of Shipping, 1995b.

Figure 11: Specifications of Private Shipyards in Turkey

	Name	Area (sq.m)	Slipways (m)	Fl'ting docks (tons)	Max. New'bg (dwt)	Capacity New'bg (ton/pa)	Capacity New'bg steel (ton/pa)	Repair steel (ton/pa)	Max lift'g/ dck'ng cap.	Repair cap. (dwt/pa)	Elect. power (KVA)
1	Gemsan	5,926		6,620			5,000	5,000	14,000	280,000	1,000
2	Hirodi-namik	17,085	120 x 25 / 120 x 25 / 95 x 20		6,500 / 6,500 / 3,500	6,500	17,000	1,000	6,500	320,000	250
3	Gemak	17,375	110 x 22 / 130 x 22	7,500	9,000 / 12,500	8,000 / 12,500	2,000 / 3,000	20,000 / 3,000	2,000	100,000	640 / 1,640
4	Desan	17,313	120 x 22		8,000	8,500	2,000	1,000	3,000	36,000	630 / 1,640
5	Sahin Celik	9,139	50 x 20 / 110 x 20		3,500 / 6,500	3,500 / 6,500	1,000 / 17,000	750			
6	Yildirim Gemi	13,649	90 x 18 / 80 x 16 / 70 x 16		6,500 / 5,000 / 2,500	10,000	2,500	1,000	2,500	60,000	900
7	Gemyat	14,669	120 x 22 / 90 x 47		10,000	10,000	2,500	10,000	3,000	36,000	1,000
8	Anadolu	40,250	110 x 30 / 110 x 26		7,500	7,500	2,000	1,500	2,500	30,000	450
9	Deniz End	40,250	135 x 20 / 117 x 30 / 100 x 20		130,000	16,000	3,500	4,000	3,000	35,000	250
10	Turkter	41,250	70 x 80		3,000	3,000	1,100	500	60		650
11	Yildiz	4,150	110 x 20		6,000	6,000	1,500	1,800		120,000	630
12	Celik Tekne	82,500	120 x 20 / 60 x 25		12,500	16,500	4,000	3,000		500,000	630
13	PKM	97,250	85 x 33		5,500	6,500	2,000		60		600

Figure 11: Continued

14	Sedef	132,312	120 x 16.5 / 120 x 20.8		20,000		15,000	2,000		600,000	2,540
15	Tuzla Gemi Endustr-isi	41,500	110 x 25 / 130 x 27.5		12,500	25,000	4,000	1,500		500,000	400
16	Selah Makina	43,000	122 x 23 / 140 x 30 / 84 x 12.9		14,000	20,000	7,000	500		300,000	650
17	Dearsan	21,500	100 x 20		6,000	6,000	1,700	1,000		30,000	1,000
18	Rota	27,467	80 x 10 / 80 x 16		3,500	3,500	1,100	1,500	1,600	20,000	650
19	Torlak	19,627	60 x 30.5 / 50 x 18.5		3,500	4,000	1,200	850	1,500	20,000	800
20	Iyideniz	15,853	115 x 24 / 50 x 16		7,500	6,000	1,800	2,000			650
21	Ceksan	15,444									
22	Gisan	14,940	100 x 32 / 110 x 22		8,500	12,000	3,500	1,500	2,500	30,000	400
23	Torgem	30,000	135 x 23 / 100 x 28 / 100 x 19	5,000	20,000	15,000	5,000	2,000	18,000	400,000	1,550
24	Celiktr-ans	16,712	120 x 16 / 100 x 14		5,000	10,000	3,000	1,500	2,500	30,000	35
25	Gemtis	4,400	150 x 17		12,000	4,300	1,500		10,000		60
26	Dortler Engin Deniz	7,650	70 x 8.5 / 80 x 8.5		3,000				1,500		77
27	Tuzla Tersane-si	24,978								432,000	

Figure 11: Continued

28	Doitel	5,164							
29	Cindem-ir	5,446				600	100,000		140,000
30	Dentas	13,850		10,000		4,000			120,000
31	SS Nuh								
32	Notika	5,000							
33	Marma-ra Transp-ort	20,000	120 x 24 / 120 x 18	14,000	18,000	4,500		PIER	1,000
34	Um Denizcil-ik	BUILDING STAGE							
35	Madenci Gemi Sanayi	7,120	125 x 32 / 110 x 16	15,000	13,000	3500			1,000

Source: Authors' survey.

Figure 12: Specifications and Capacities of State-owned Shipyards in Turkey

Region	Shipyard	Area (sq m)	Slipways (m)	Total Employees	Engineers	Max Newb'g capacity (dwt)	Newb'g capacity steel (dwt/year)	Newbuilding capacity steel (ton/year)
Istanbul Halic	Halic	69,810	56 x 18 90 x 22	727	37	5,800	11,100	3,169
Istanbul Halic	Camialti	72,000	140 x 24 92 x 16.5	496	35	18,000	20,800	5,934
Istanbul Pendik	Pendik	953,000	202 x 38 300 x 70	1,284	90	17,000	143,000	31,790
Izmir Karsiyaka	Alaybey	71,433		243	18	1,000	3,000	1,184
General management				210	38			
Total				2,960	218		177,900	42,077

Source: Government of Turkey, 1995.

Long-term loans from the Government for the shipbuilding industry in Turkey are not available as shown in Figure 12. However, there have been various long and short-term loans for Turkish ship construction from other countries from time to time. For example, the most recent long-term loan for the Turkish shipbuilding industry has come from a Japanese loan. The agreement for this loan, totalling US$ 250 million, was signed in March 1995 and was planned to be used from the beginning of summer 1996 (Chamber of Shipping, 1995c). The main targets were to newbuild approximately 300-350,000 dwt of ships in Turkish shipyards and to modernise Turkish shipyards with Japanese technology (Drewry, 1995).

Privatisation in the Turkish Shipbuilding Industry

After Mrs Ciller was elected as the Prime Minister from the True Path Party in June 1993, one of the targets of the Government was to take further steps towards privatisation of various state owned companies from a number of sectors. The main target was to convert the state owned companies, which had been heavily subsidised by the Government, into profitable private companies and one of the major sectors for these privatisation plans was that of the maritime industry (Lloyd's Ship Manager, 1993). The plans for maritime privatisation focused upon state owned ports, shipyards and the Turkish Cargo Lines Company that operated a large variety of cargo vessels.

It was soon realised that the legal problems of privatisation could only be overcome by minimising the bureaucratic delays that characterised Government planning and procedures. The whole process was given additional impetus by the Prime Minister who had a major policy of economic liberalisation and transformation of the Turkish economy as a whole.

Additionally, privatisation was also highly recommended by the Chamber of Shipping which emphasised that state owned ports and shipyards should be privatised immediately. The President stated that the Chamber believed that the state cannot conduct trading, and therefore, state owned ports and shipyards needed to be transferred to the private sector (Lloyd's Ship Manager, 1993).

One of the state owned maritime companies, Turkish Cargo Lines, was also earmarked by the Government for privatisation. Since some of the activities of this cargo vessel operator had been unprofitable, it was difficult to find buyers from the private sector (Lloyd's Ship Manager, 1994). Meanwhile the President and Chairman of the Chamber of Shipping, continued to make calls to the Government for quicker privatisation actions in the maritime sector stating that there was no further time to lose in enacting legislation.

Unfortunately, as a consequence of the severe bureaucracy that remained in Turkey by 1995, many problems still existed. To quote the President of the Chamber (Lloyd's Ship Manager, 1995):

> "The public authorities are acting against the private sector because of the severe bureaucracy in Turkey. Some of the state owned companies are growing older and worse. In my opinion, privatisation has not been realised seriously, and remains as the politicians' promise. So far not a single state owned company in the maritime sector has been privatised. The most important matter in Turkish shipping, at the moment, is the privatisation of the state owned companies, which should happen immediately covering all sectors of maritime, including ports, shipyards and pilotage. Turkey is trying to adapt to integration with the European Customs Union, therefore, we believe that the existing obstacles should be removed and we should do our best to go further."

Problems with the Turkish Shipbuilding Industry

The primary prerequisite for the shipbuilding industry to be successful is the selection of the right markets and the utilisation of proper marketing methods that are suitable for those selected markets. Receiving orders and achieving successful deliveries depends upon financing. The scope of financing includes both short-term finances of the shipyards and mid-term finances of the shipowners.

The conditions in Turkey in the mid 1990s did not comply with the above. In addition to these failings, foreign exchange rates applied by the Turkish Central Bank and by the Treasury as well as wage policies followed within the public sector, eliminated the possibility of price advantage for the Turkish shipbuilding industry. Moreover, the situation continued to worsen and the contribution of this sector to the development of the country in general and to foreign trade, deteriorated due to the lack of resources.

The main problems of the Turkish shipbuilding industry by 1995 can be listed as follows (State Planning Organisation, 1995; Chamber of Shipping, 1995a):

- problems due to the general political context for the shipbuilding industry;
- declining demands for new shipbuilding;
- increasing overcapacity within yards due to the declining number of newbuildings;
- ship repair problems;
- requirements of improving shipbuilding technology;
- lack of financial support for the industry;
- the significant rise in the costs and prices of newbuildings due to the high inflation rate;
- the shortage of loans and subsidies;
- very slow action taken by the Government in privatising state owned shipyards;
- absence of marketing techniques for shipyards.

Conclusions

The shipbuilding industry has an important role in the Turkish maritime sector in terms of its contribution to the general economy. The annual production capacity of the Turkish shipbuilding industry with its present production techniques and technology is approximately 488,000 dwt or 130,000 tons of steel processing in a year. Additionally, the annual ship repair capacity of the industry is approximately 7 million dwt at present. With the enlargement of capacity at Um Shipyard, the total capacity of the shipyards will reach approximately 730,000 dwt per year. In order to use this total capacity and decrease the amount of overcapacity, certain steps need to be taken. The most important ones were already noted in the Seventh Five-Year Development Plan for the period between 1995-1999 but remain unfulfilled in many cases:

- privatisation actions should be speeded-up by the Government;
- technology together with the substructures of the shipyards should be improved;

- subsidy by the Government should be re-activated and speeded-up.

Regarding the Seventh Five-Year Development Plan for the period between 1995-1999, the following is a summary of the recommendations and the necessary actions for the shipbuilding industry. The annual production capacity of the Turkish shipbuilding industry with its production techniques and technology is found to be around 480,000 dwt whilst the annual ship repair capacity of the same is nearly 7 million dwt. The utilisation rate of the present newbuilding capacity is under 20% and when the existing orders are analysed, this rate will decrease to under 10% if the necessary precautions are not taken.

As previously indicated, the shipbuilding sector is highly dependent upon the secondary sector as an auxiliary industry. Therefore, the volume of employment it generates cannot be limited to the number of workers employed directly in the shipyards. Nevertheless, considering the rate of domestic contribution at nearly 40-50%, it is fair to conclude that every employee in the shipbuilding industry in Turkey creates employment for around three people in the secondary shipbuilding industry.

In the light of these findings, we can reach the following conclusions:

- foreign exchange substitute: US$ 1.2 billion/year;
- net amount of prospective foreign exchange input: (the difference between export and import values) US$ 600 million/year;
- volume of employment : 105 million man-hours/year;
- state revenues (employee's income taxes, premiums and other deductions plus the interest of foreign exchange substitutes): US$ 340 million/year;
- another fact to be considered is that with the implementation of the proposed technological measures, there will be at least 20% increase in the current capacity.

Therefore, the achievements from reorganisation and other incentives introduced for the industry during the Seventh Five-Year Development Plan should be as follows:

- new building capacity: 1,000,000 dwt/year;
- repair capacity : 7,500,000 dwt/year;

- foreign exchange substitutes: US$ 1.5 billion/year;
- employment (secondary shipbuilding industry included): 125 million man-hours/year;
- domestic contribution: 65%;
- realisation of these targets will only be possible through the implementation of the measures proposed as a whole.

The measures that need to be taken by the industry are summarised below:

- maritime training and research; recognising the need to train skilled technicians and researchers;
- measures to support projects and marketing including the export and repair of ships;
- measures to support investments in infrastructure - converting shipyard areas into organised industrial regions;
- incentives; investment in shipyards should be included within the scope of 100% investment discounts and they should benefit from incentive premiums and subsidised loans; the import of all kinds of machinery and equipment should be exempt from all tax and fund charges; the products of the secondary shipbuilding industry used in the building and repair of ships should be treated as exports; the project mortgage and loan risks insurance should be accepted as warrants against the loans provided within the range of incentives; Value Added Tax should not be imposed on imported machinery and equipment for shipbuilding and ship repair; fund sourced loans should be paid on a foreign exchange basis.

Measures to be taken by the industry should include:

- the immediate preparation of investment plans together with an assessment of financial requirements of the shipyards;
- the establishment of modern and planned production techniques;
- the establishment of a quality administration concept in compliance with ISO-9000, documented by a certificate system;
- precautions by the shipyards for the protection of the environment in both organisational and individual terms.

References

Chamber of Naval Architects (1989) *1st National Maritime Congress - Report on Naval Architecture Sector*, Istanbul, Turkey, 6.

Chamber of Naval Architects (1995) *Proceedings of Naval Architecture Congress - 1995*, Istanbul, Turkey, 3-8.

Chamber of Shipping (1995a) *Report on Turkish Shipping Sector - 1994*, Istanbul, Turkey, 198, 244-246.

Chamber of Shipping (1995b) *Turkish and World Shipping in 1995*, Istanbul, 16-18.

Chamber of Shipping (1995c) *Turkish Shipping World*, March-April, Istanbul, Turkey, 10.

Chamber of Shipping (1996a) *Turkish Shipping World*, February, Istanbul, Turkey, 9-18.

Chamber of Shipping (1996b) *Turkish Shipping World*, March, Istanbul, Turkey, 12.

Drewry Shipping Consultants (1995) *The Shipbuilding Market*, London, 37, 38, 69.

Government of Turkey (1995) *Sectoral Report on Turkish Shipping 1994*, Istanbul, Turkey.

Interview with Mr Kenan Torlak, Head of Turkish Shipbuilding Association, April 6th, 1996.

Lloyd's List (1993) *Turkey*, October, Lloyd's of London Press Ltd., London, 13-14.

Lloyd's Register of Shipping (1994) Istanbul, Turkey.

Lloyd's Ship Manager (1993) *Turkish Shipping and Ports Directory*, Lloyd's of London Press Ltd., London, 7, 17.

Lloyd's Ship Manager (1994) *Turkish Shipping and Ports Directory*, Lloyd's of London Press Ltd., London, 9, 23.

Lloyd's Ship Manager (1995) *Turkish Shipping and Ports Directory*, Lloyd's of London Press Ltd., London, 9.

Ministry of State (1992) *Towards Ministry of Maritime*, Ankara, Turkey, 15.

Nehir, L. (1993) *Integration of Ship Industry and Defense Industry*, Undersecretaries of Defense Industry, Ankara, Turkey, 10.

Shipping World, September-October, Istanbul, Turkey, 14.

State Planning Organisation (1995) *Shipbuilding Industry*, Ankara, Turkey, 6, 18.

Tez, I. (1993) *Discussions on Turkish shipping*, Ankara, Turkey, 6.

UNCTAD (1994) *Review of maritime transport - 1993*, UNCTAD, New York, 3-52.

10 Reliability, Availability, Maintainability and Safety Assessment in the Maritime Industry

BURAK ACAR
SCHOOL OF MARITIME BUSINESS AND MANAGEMENT,
DOKUZ EYLUL UNIVERSITY, IZMIR

Introduction

The International Safety Management (ISM) Code has been hailed by some as the ultimate legislative tool to eradicate sub-standard shipping from oceans. This will allow the quality operators to compete on a level playing field offering quality services to their customers with comparable high quality attention to safety and the environment. Classification Societies have also played an important role in marine safety. A new safety culture, migration to risk-based classification and advances in information technologies are having a profound impact upon merchant ship operations.

This paper takes a Turkish perspective in reminding readers of the rules within the IMO Safety Codes and provides information about rapid changes in the reliability, availability, maintainability and safety assessments in the marine industry.

The purpose of the International Safety Management (ISM) Code is to provide an international standard for the safe management and operation of ships and for pollution prevention. The tenth element of the Code is concerned with the 'Maintenance of the Ship and Equipment' and states amongst other things that the company should establish procedures to ensure that the ship is maintained in conformity with the provisions of the relevant rules and regulations and with any additional requirements which may be established by the company.

The company should ensure that:

- inspections are held at appropriate intervals;
- any non-conformity is reported, with its possible cause, if known;
- appropriate corrective action is taken; and
- records of these activities are maintained.

The company should establish procedures in its Safety Management System (SMS) to identify equipment and technical systems, the sudden operational failure of which may result in hazardous situations. The SMS should provide for specific measures aimed at promoting the reliability of such equipment or systems. These measures should include the regular testing of stand-by arrangements and equipment or technical systems that are not in continious use. The inspections should be integrated into the ship's operational maintenance routine.

Reliability is often overlooked during early developmental testing when the focus is on improving system performance. As a result, rapid reliability growth programmes are frequently initiated late in development, just before the system enters production.

Initially, in the ship design process, much of this data is not fully defined. At that stage, we rely on approximate data based on preliminary design and previous ship experience. The analysis continues until the final stages of the design thus providing the ship designer with useful guidance on effective ship survivability features.

Ship systems are designed to meet availability requirements stated in ship specifications. To achieve that level of performance the system's component availability is determined from manufacturer-supplied information or from data banks.

Reliability of a ship component is a function of its material, design, and manufacturing quality, age, elapsed time since maintained, and operating environment. The predicted time of component failure is defined in most simulation models as a probability distribution function that includes a Mean Time Between Failures (MTBF) provided by the component manufacturer. Recent models have more sophisticated functions which account for the effects of inspection and failure discovery as well as differences in full-time operation in contrast to use on demand or standby. Equipment reliability information must be collected and maintaied to assess the effects of engineering and environmental changes and ship vulnerability.

Maintainability of equipment is a function of its design quality, availability of parts and labour and ability to repair the equipment within the given mission scenario. Similar to the time of failure, the time to repair, or down time, is defined as a Mean Time to Repair (MTTR) statistical function. MTTR data are available from equipment manufacturers' sources or from the data banks.

Dependability Assessment in Design

Predictions cover both ship machinery and a ship structure's useful life; their aim is not to forecast the probability of early failures. Predicted reliability should be compared only with operational reliability after failures have been eliminated, that is after pre-operational tests have been performed. The better the qualification and pre-operational tests are designed and carried out, the more accurate the predictions will be.

Dependability assessment studies should motivate its enhancement. There is a need for a real dependability programme if dependability assessment is to be optimised. This programme must provide the necessary organisational structure, responsibilities, procedures, activities, expertise and means which will ensure that a system will satisfy the dependability requirements of a particular project.

Likewise a policy of dependability assurance must be implemented to guarantee that the required goals will be attained and that dependability will be what it is supposed to be. This assurance policy is enforced right from the design stage and must be maintained during operation.

Safety Culture and Availability

The reliability and risk based approach for safe and reliable operation has been successfully used in other industries. At a very late stage, the merchant shipping industry is beginning to take advantage of reliability engineering. New initiatives from the IMO, such as the ISO and STCW codes and introduction of formal safety assessment guidelines, are changing the management of safety in ship operations.

Classification societies have started introducing risk-based rules for ship machinery in safety and operational efficiency and some companies and projects focus on the implementation of quantitative reliability and risk assessment. Application of reliability centred maintenance techniques to merchant ships has been proposed and a number of pilot studies have been carried out. Formation of an industry-wide ship machinery reliability database was first proposed in the USA in the 1960s. In 1965, The Society of Naval Architects and Marine Engineers organised Panel M-22 for Reliability and Maintainability. This panel initiated two major tasks:

- preparation of a practical guide in reliability and maintainability, and
- development of a practical shipboard data reporting system for a data bank.

Organisation and Management of Dependability Assessment

Specialists, who design and compute operation systems, often have difficulties in admitting the occurrence of failures and will tend to look upon failures as nonsensical.

Dependability specialists must have an insatiable thirst for technological knowledge, must be capable of rapidly assimilating a new technology, have a real power of abstraction and a good morale to stand constantly dwelling in a world of failures. Experience is essential in this field to avoid two extreme attitudes:

- Quantitative assessment is useless or not credible.
- Figures are indisputable. Therefore the specialists hang on to them as though their life depends on them.

Dependability assessment may structure the various related fields of activity: new working and communication networks may emerge within the company. This may contribute to the birth of a true dependability culture.

Generally, dependability assessment is valid only at a given time in its life cycle. Gradual change of a project during the design phase and later during the operating phase makes updating the dependability assessment necessary; this may have an impact on the methods adopted and encourage the use of computer programmes to facilitate assessment updating.

Assessing Safety

The problem of safety becomes more urgent when we consider the fact that industrial activities in the marine industry have grown both in number and scope thus giving birth to new hazards. Right from the design stage, it is essential to perform risk assessments. This approach raises several questions including that of the acceptable level of risk.

The safe, efficient operation of ships is everyone's concern and the primary objective of the majority of vessel operators worldwide. If a rational and quantitative approach to safety is to be developed, the following questions must be considered:

- What are the risks generally faced by the shipowners/shipping companies?
- What studies have been performed for hull and machinery development?
- What are the safety goals set in other companies?
- How often are new technologies used on ships?

A situation becomes a hazard if it can be harmful to people, society or the environment. The occurrence of an undesirable event is generally measured by its occurrence probability over a given period. The undesirable event can have various effects:

- Consequences on humans: injuries or loss of lives.
- Economic consequences: cost of repair, operation loss, etc.
- Environmental effects: air and sea pollution, injured or dead animals.

The risk factor must also be considered. Tables 1 and 2 give details of some particularly serious catastrophes. It is sometimes more meaningful to calculate the risks per hour of exposure and per exposed individual.

Human Factors

In deriving the generic ship model that forms the basis of safety assessment it becomes apparent that human factors dominate many of the risk scenarios. Casualty data confirms the validity of the general assumptions, which often suggest that around 80% of casualties result directly from human error. It is clear to any safety analyst that it is not possible to separate technical safety from the influence of operators, in the very widest sense.

Table 1: Catastrophic Natural Hazards

Hazards	Estimated frequency	Number of fatalities
Tornado	37.5 over 100 years	137-250,000
Hurricane	37.3 over 100 years	200-900,000
Landslides	6.74 over 100 years	400-4,000

Table 2: Catastrophes Caused by Man

Hazards	Estimated frequency	Number of fatalities
Ship disaster	More than 25 over 30 years	17-1,913
Accidents from motor vehicles	Very sizeable number	250,000 fatalities and 7,500,000 injured per year

The introduction of the ISM Code has provided greater opportunity for the classification societies to become closely involved with human factors. Lloyd's Register (LR) has taken additional steps by becoming active in promoting the assessment of crew training establishments, required by STCW.

Management of Risk and Risk Targets

A safety regime that encompasses safety assessment, would probably encourage wider application of risk based methods for ship management, promoting the use of decision making on the basis of risk analysis. A modelling package, as shown in Figure 1 including the fleet system can be used. In other industries, it is quite normal to carry out project viability studies on a risk basis to determine whether there is an acceptable probability that business objectives will be attained. In marine terms for example, it is possible to use probabilistic modelling to assess the delivery rate of LNG between two terminals with a given fleet capacity in probabilistic terms. Similarly, the probability of a cruise ship achieving an acceptable level of destination arrivals can be estimated. Whilst not an exact science, the results can provide considerable assistance is defining business perspectives. The same methods can be used to predict the likelihood of major casualties and to feed the information into the risk management systems.

Figure 1: Outline of the Fleet Modelling System

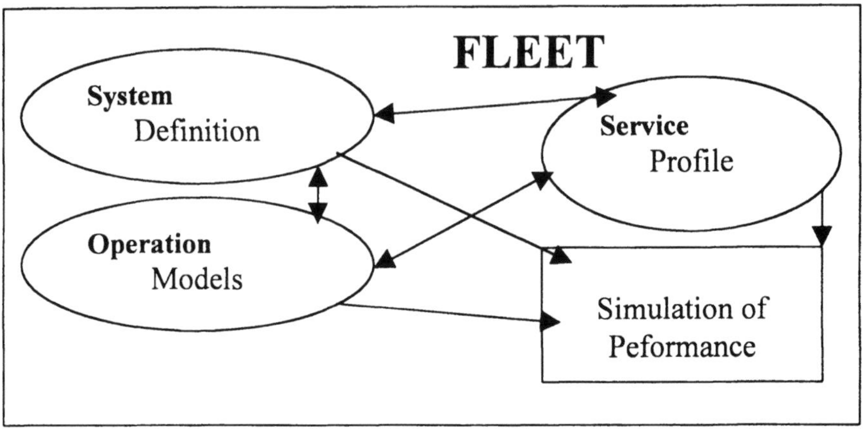

Data Analysis

Trends can be established and analysed to improve shipping operations in three ways. First, meaningful comparisons can be made between similar equipment types in order to make the most effective renewal and replacement decisions. Second, maintenance resources will be more efficiently used through the practical migration from a time based, to a reliability based maintenance system.

According to many business plans, customised data reports will be available to any organisation for a nominal fee. A not-for-profit operation organisation is the best structure to accomplish this goal.

By utilising data, ship owners can develop new mechanical and structural designs, which will be optimised from an initial cost as well as a life cycle cost basis. The resultant ship designs should enjoy higher levels of reliability, availability and maintainability, while potentially minimising the initial capital expenditure and reducing operating costs.

Conclusions and Recommendations

Reliability analysis is a logical and necessary addition to ship vulnerability assessment processes. This safety assessment process provides vessel equipment performance information in a suitable format that allows shore-side and shipboard maintenance management personnel to better target their available resources.

The adoption of the safety assessment by IMO has introduced a new dimension to the way that safety is considered within the shipping community. It is expected that the adoption of safety assessment, together with the introduction of changes to SOLAS and STCW, will advance the application of risk based thinking in the marine industries.

Regardless of all the difficulties, ISM is essential to the future of the industry and owners ignore it at their peril. However, no law will work without the willing participation of the majority of those affected. Without adequate policing and enforcement, which tries to ensure the intent of the legislation, the Code will not succeed.

References

Aldwinckle, D.S. and Pomeroy, R.V. (1984) A Rational Assessment of Ship Safety and Reliability, *Transactions of Royal Institution of Naval Architects*, RINA: London.

Inozu, B. and Radovic, I. (1999) Practical Implementation of Shared Reliability and Maintainability Databases on Ship Machinery: Challenges and Rewards, *Transaction of Institute of Marine Engineers*, December.

Inozu, B., Schaedel, P.G. and Molinary, V. (1996) *Reliability Data Collection for Ship Machinery*, Paper Presented at the Meeting of the Gulf Section of the Society of Naval Architects and Marine Engineers.

Malakhoff, A., Klinkhamer, D. and McKesson, C.B. (1997) *Analysis of the Impact of Reliability, Availability and Maintainability on Ship Survivability*, John J. McMullen Associates, Inc.

Pomeroy, R.V. (1987) *The Generation and Use of Reliability Data for Marine Engineering Systems*, Proceedings of ICMES' 87.

Pomeroy, R.V. (1999) *Classification - Adapting and Evolving to Meet the Challenges of the New Safety Culture*, Proceedings of the Conference of New Safety Culture, Part II, The Institute of Marine Engineers: London.

Umimoto, T. (1998) *Guidelines to International Safety Management (ISM) Code Certification*, Lloyd's Register of Shipping: London.

Villemur, A. (1992) *Methods and Techniques, Volume 2, Reliability, Availability, Maintainability and Safety Assessment*, John Wiley and Sons Ltd: London.

11 Advances in Ergonomics in the Maritime Sector: a Turkish Perspective

GÜLER BILEN ALKAN AND OSMAN ASKIN BAK
DOKUZ EYLUL UNIVERSITY, IZMIR

Introduction

Ergonomics can be simply defined as the study of work, and adapting the job to meet the worker's needs to reduce or eliminate injuries or illnesses.

Employees are valuable and replacing an employee who is injured drains money, morale and talent from an organisation. In addition, companies are able to demonstrate that implementing ergonomic standards has reduced the incidence of musculoskeletal disorders, decreased lost workdays, significantly lowered workers' compensation costs, and improved performance, productivity and quality.

Ergonomics in shipping is a new area where ergonomists or shipping consultants have applied new approaches to the ship design stage through to the operation of the ship. Education plays a very important role in this approach.

Substandard ships threaten the oceans with pollution risks. Substandard working conditions threaten the life of the crew, passengers and cargo. Working in shipping is a difficult profession. Ergonomic design of the ship and workplace increases efficiency and decreases the risk of injury and loss of life.

In this paper, ergonomic studies in shipping and maritime transport are investigated through an examination of the literature. An ergonomic approach to work places and working conditions in shipping is also analysed. A survey has been conducted amongst shipping companies and the results that have been derived are briefly analysed.

The world is not what it used to be. Neither are ships, shipowning companies, ship managers, insurance companies or even Classification Societies. Each lives in a changing world with new and serious challenges which need new modes of operation. The attitudes and impact of environmental forces upon society have also changed and people are more concerned about the human and technological impacts upon the environment.

The awareness of global pollution threats has created a new consciousness and is the common concern of everyone. The general public will no longer excuse nor accept failures in the human factor. This clearly means that lack of sufficient competence and poor attitudes to safety for both shore management and ship crew must not remain as the main cause of loss of life or serious environmental pollution.

This new attitude is equally valid whether it relates to shore-based activities, heavy or process industries, or to offshore activities or sea transport.

The focus on the human element when disasters occur makes it necessary, by means of new approaches, to achieve reduction of risks and further control of losses. The relationship between the human factor and maritime hardware brings new challenges into safety work. Hence increased attention is paid to the fact that more than 80% of all undesired events can be referred back to human error or substandard acts or practices.

As we progress through the 1990s and into the new millennium, the application of ergonomics to ship design will become paramount particularly in the interest of the crew. A good example is bridge design and the increasing emphasis upon unmanned machinery spaces. This involves the concept of a ship control centre which embraces navigation and engine and cargo control functions. A further example of influence of ergonomics involves ship systems concerned with how people get aboard and leave ships and how they move around the ship once aboard. This involves stairways, lifts and hoists; machinery spaces; ladders - fixed and portable; mast and radar platforms; lifeboats; holds and tanks; emergency escape facilities; watertight doors; stores and workshops; and sick bay and hospital rooms.

Ergonomics can play an influential role in the panel instrument and display layout and configuration both in the bridge/wheelhouse and machinery room. This involves the choice of hardware and its disposition, the design of alarm and warning signals, and the overall philosophy of both the control and maintenance panel layout. The influence of ergonomics may extend to the written word in the organisation and preparation of manuals, check lists and fault-finding guides and in the provision of on-board computer-aided systems which run on an interactive basis.

The modern engine room is computer-controlled and electrically operated through a series of panels on the bridge. Undoubtedly, in the future, further emphasis will be placed upon computer technology in ship design and operation.

Objectives and Methodology

Human beings play an important role in the functioning of socio-economic systems. Ergonomics or human factors is an interdisciplinary science that deals with the study of the interaction between human beings, the objects they use and the environments in which they function. In the literature, the first use of the term seems to have occurred during the mid-nineteenth century. A Polish educator and scientist, Wojciech Jastrzebowski, introduced the term combining two Greek words: *ergon* meaning work, and *nomos* meaning laws (Pulat, 1996).

Design for human use is another practical definition of ergonomics. The central theme of ergonomics is *fitting the task to the person*. This means that designers must design for human use, keeping in mind at all times what people can and will do.

There are three major objectives of ergonomics:

- Comfort;
- Well-being;
- Efficiency - physical, mental and productive.

The most important assumption made within ergonomics is that equipment, objects and environmental characteristics influence human performance and thus the total human-object system performance.

Frequently, we hear that human error was the cause of a major accident. However, detailed investigations reveal that the real reasons are often design deficiencies that led to the human error.

A model relating to Human Integrated Design is shown in Figure 1. This model should be considered in the ship planning and design stages. Operational variables (input from ship operators), hardware (the technical elements of the ship) and other systems (quality systems, international regulations, economic systems) must be considered in terms of their impact upon the human role. The outcome of these systematic approaches should be control of the design process by a series of design rules.

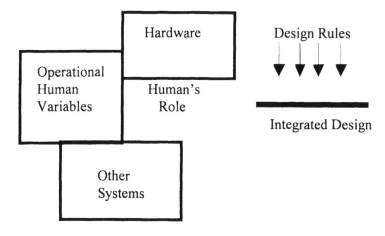

Figure 1: The Various Components of Human Integrated Design

The Cost of Ignoring Ergonomics

These are both numerous and serious and include:

Less productive output;
Increase in lost time;
Higher medical costs;
Higher material costs;
Increased absenteeism;
Low quality work;
Injuries, strains etc;
Increased probability of accidents and errors;
Increased labour turnover;
Less spare capacity to deal with emergencies.

Potential gains from introducing the concept of ergonomics to the design process are the reverse of the above.

The Ergonomic Design of the Ship

Shipping is a very hazardous industry with a harsh environment and numerous safety and industrial hygiene problems. It presents some unique challenges.

Operational design is that part of the design process that ensures that a ship can be worked effectively by the crew and, where appropriate, stevedores. It covers the concepts of the ship at sea, controllability, workability, manoeuvrability, security, safety and emergency response, maintainability and habitability (Mol, 1992).

Accommodation standards are generally improving and this should be recognised when planning for a new ship. However, specific issues sometimes emerge that need to be considered – for example, Korean and Japanese ship designs often assume that the height of seafarers will be 5' 6" rather than 6' so extra length bunks and settees may be required for European crews.

Generally bunks should be facing fore and aft with seats across the ship. All cabin fixtures and fittings must be capable of being secured against ship movement. The sleeping pattern of watch-keepers must be assessed and their cabins segregated from other cabins or washroom areas where other personnel may disturb them during their rest period.

Noisy equipment, which may be worked at night, should not be fitted directly adjacent to accommodation. Vibration levels must be assessed and avoided in accommodation areas. Noise levels on board should be in accordance with safe long term limits. Noise dampers should be fitted around noisy equipment. The siting of the engine exhausts, air conditioning intakes and galley extractors must be carefully considered using models or wind tunnel tests to establish the optimum layout for predominant prevailing conditions.

Ergonomic Programmes

Ergonomics can be implemented into shipping by means of programmes prepared by specialist consultants or ergonomists. The programme must include the following items.

Leadership

There should be one person clearly in charge. Commonly, when a company wants to introduce a new programme or process, they send everybody for training. The idea behind the training or education programme is that on return, new procedures and concepts should be put into effect by staff and this is likely only to occur if there are clear, identifiable leaders who can take charge and direct the changes. The new plans and procedures need to be written and then communicated to all those who need to be aware of the changes (OSHA *et al.*, 1997).

Management Commitment

A management commitment is essential and this means commitment to devote resources, time and money to solving ergonomic related problems.

Education

Education is the key to success in an ergonomics programme to help individuals to understand the nature of these problems and that they are real and controllable. Employees need some form of training to define ergonomic issues, define the risk factors and to indicate the chain of responsibility.

Employee Involvement

This is also critical and if it involves a team work approach then employees can become part of the decision making process. Employees also know more than anyone about the ergonomic issues that characterise their employment.

Job Modification

Job modification is much more likely to succeed if worker input is encouraged.

Some other factors that ergonomics should consider and can help to improve working conditions include:

- Noise;
- Vibration;
- Displays;

- Illumination;
- Indoor Climate;
- Outdoor Operations;
- Radiation.

Human Error

It has been said that 70-80% of marine accidents are caused by the human factor. An accident may or may not lead to physical harm to an individual; however, no matter the outcome, the fundamental question to be answered is: how do accidents happen? A number of different causes clearly exist:

- unsafe acts and/or unsafe conditions;
- situational factors (including psychological and physical characteristics of the environment and the workers);
- specific issues such as – temperature; insufficient light; length of the working day; layoff rate etc.

The most effective method of controlling human errors is the ergonomic design of work. Most of the causes of human error are related to conflict between the user and a hostile work design. Human beings make fewer errors in environments that are compatible with their expectations and in which they function effectively. Another approach in the past has been replacing the operator. The assumption of this approach is that poor performance is due to personnel factors such as deficient dexterity, poor vision or inadequate skills. A more proficient operator will make fewer errors. Finally, systems can be designed with human error in mind. However, it should be remembered that is not possible to eliminate human errors totally.

Risk Factors in Shipping

There are many of these – perhaps more than most industries, and they stem from a variety of circumstances including:

- Work practices;
- Individual factors;

- Ergonomics, human/machine interface;
- Technical elements;
- Management failings;
- Environmental conditions;
- Infrastructure;
- Working conditions;
- Terminal conditions;
- Other external factors.

Figure 2: Risk Factor Structure

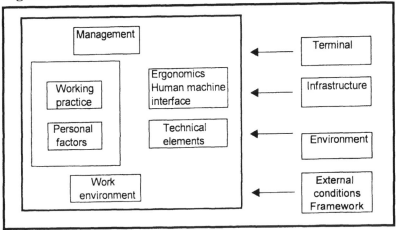

RISK FACTOR STRUCTURE

Maritime transport system itself has both hard and soft features. The hard features consist of the technical systems and man/machine interface. The soft has two basic elements: the human individual and the management system. The human component may further be broken down into personnel factors and observed working practice (Kristiansen, 1992).

Risk factors in shipping can be evaluated in terms of a systems approach. Scientists and ergonomists need to research the human element of risk factors to be able decrease their effects in shipping. A review of Turkish maritime related companies including shipyards, shipowners and manning agencies showed that the concepts of ergonomics were little known in this sector. The principles of ergonomics and ergonomic programmes need to be introduced into maritime companies through an extensive education programme.

The annual cost of marine accidents to the maritime sector is estimated to be at least US$10 billion. Any improvement by using an ergonomic approach will save millions of dollars for the sector. We have to use every tool that deals with the relationship of humans and their work environment in an attempt to decrease human errors.

Conclusion

The cost of ignoring ergonomics is very expensive in terms of low productivity, lost man-hours, workers' compensation and efficiency in the maritime sector. Ergonomics should not be only the work of industrial engineers, occupational health experts and human factors specialists. Operation managers, ship designers, naval architects, captains and shipowners amongst others, should become more involved with this approach to human factors. The benefits from this will fall to all of us. The maritime sector has some effective tools including the ISM Code and the STCW Convention both of which are mandatory and require that ergonomic issues are considered. Maritime universities and colleges need to consider these issues more closely and incorporate them within their curricula. The ISM Code came into force to eliminate substandard ships; ergonomics programmes should be used to prevent substandard working conditions.

References

Institute of Industrial Engineers (1985) *Industrial Ergonomics: A Practitioner's Guide*, IIE: London.

Institute of Industrial Engineers (1990) *Industrial Ergonomics: Case Studies*, IIE: London.

Kristiansen, S. and Rensvik, E. (1992) *Human and Organisational Factors in Safe Operation and Pollution Prevention*, IMAS 92, Quality of Shipping in the Year 2000, 11-13 November.

Mol, J. (1992) *The Human Factor in Ship Handling and its Effects on Marine Safety and Environment*, First Joint Conference on Marine Safety and Environment Ship Production, Delft University of Technology, 1-5 June.

OSHA and National Institute for Occupational Safety and Health Ergonomics (1997) *Effective Workplace Practices and Programs*, (NIOSH), Chicago, January 8-9.

Pulat, B. M. (1996) *Fundamentals of Industrial Ergonomics*, School of Industrial Engineering, University of Oklahoma, Waveland Press Inc.

12 Liner Agents and Container Port Service Quality[1]

D. ALI DEVECI, A. GÜLDEM CERIT
DOKUZ EYLÜL UNIVERSITY, IZMIR
AND JACQUES H.B. SIGURA
TRANSMAR SHIPPING AGENCY CO. LTD., IZMIR

Introduction

With the development of containerisation and intermodalism, ports have changed their conventional role to one of a dynamic node in the complex production distribution network. The physical, commercial, managerial, industrial and logistical services provided to port users have increased in order to obtain more traffic and to meet the developing requirements of international transportation and trade. Ports are service producing businesses and the service characteristics and quality of a container port are of vital importance to both the shippers and the shipowners together with the liner agents as the representative of the latter.

This study analyses the service characteristics of a port with respect to views of the liner agents in container transportation. An empirical study is undertaken through a questionnaire applied to liner agents situated in Izmir and results are directed towards service characteristics and service quality of the Port of Izmir.

Whatever the requirements of the users are, the trend is towards higher port service quality and higher performance through contemporary approaches. The developments in international trade directly affect shipping and ports in particular as nodal points in international trade and transport and thus adaptation is needed to these new requirements.

---

[1] This paper was originally presented at the IAME Halifax Conference in September 1999.

Port Services and International Quality Requirements

The multiplicity of world trade centres calls for an extensive transport network and a greater variety of transport services should be provided to link the whole trade complex with ports and quality services. The increased amount of semi-finished or manufactured cargo carried by containers, requires substantial improvement in both speed and security whilst reliability (in terms of both the time and frequency of transport services) is of even greater importance. At the same time, good information and communication services are essential. Meanwhile, it is not merely the cost of transportation for a particular transport mode, but also the total costs of integration of transport and distribution that are of great importance in international trade, and ports are points where different types of transport modes meet.

Depending on the new requirements of the pattern of world trade, the concepts and practices regarding transportation/distribution are bound to change. These new concepts and practices in international transport require quality service at ports summarised by the Japan International Cooperation Agency (1996) and Peters (1989) as:

- Integration of foreign trade and the transportation chain;
- Containerisation and intermodalism;
- Logistics of the transportation chain;
- Transhipment;
- Specialisation and economies of scale of ocean going vessels;
- Inter-port competition;
- Strategies of container ship lines (use of load centres, hub ports etc);
- Customer oriented diversity and flexibility.

Recent Trends in Ocean Container Transportation and Ports

Emergence of new container shipping lines in the market, such as Asian operators and non-market economies in the 1980s and other reasons such as excessive loans for new vessels, caused excess capacity and competition among container lines (Chadwin *et al*, 1990). Faced with excess capacity, a declining freight market and fierce competition, the shipping lines responded in different ways. These changes and trends in liner shipping can be categorised under the following headings (Slack *et al,* 1996; Chadwin *et al,* 1990; Japan International Cooperation Agency, 1996; Peters, 1989):

- *Changes in liner service organisation.* Rationalising services often through collaboration with other lines; strategic alliances which may have such forms such as traditional mergers, joint operations, slot sharing agreements, shipping pools, shipping consortia as well as conferences;
- *New services.* Round-the-world services (initiated by Evergreen), pendulum services (initiated by Nedlloyd), integrated global network (initiated by Maersk);
- *New markets.*
- *Larger container ships* - in order to realise economies of scale;
- *The trunk line concept and feeder networks;*
- *Mega-carriers.*

Port Services in Ocean Container Transportation

Both the ports and terminals felt the impact of the commercial, technological, and regulatory trends of the 1980s through the expressed needs of their primary customers, the ocean carriers. The overall effect of these trends was to reinforce the turbulence and competitive pressures felt by the carriers and drive them to search for new cost, price and service strategies that offered the prospect of survival and profitability (Chadwin *et al,* 1990). The carriers, in turn, transmitted many of these pressures to the ports and marine terminals. Time and cost efficiency of port calls became critical in ocean container transportation.

Ports used in container transportation are called *third generation* ports. These ports emerged in the 1980s, principally due to worldwide large-scale containerisation and intermodalism combined with the growing requirements of international trade as described above.

The activities and services in such ports are specialised, variable and integrated. They are subdivided into four major different categories:

- Traditional port and terminal services;
- Industrial environmental services;

- Administrative/commercial services;
- Logistics/distribution services (Japan International Cooperation Agency, 1996).

Traditional port and terminal services in container transportation are carried out in container terminals. Every marine container terminal performs four basic functions and provides services to users. These services are receiving, storage, staging and loading. Containers arriving at the terminal are the 'raw materials'. Although most containers undergo no physical transformation, terminal workers and equipment perform various operations on the containers, which give them added value (Chadwin *et al*, 1990).

Facilities and equipment in container terminals serving container shipping lines are as follows (OCDI, 1990) - berth, gantry crane, container yard (marshalling yards, storage yards, CFS etc.), gate, maintenance shop, control tower, office building, other miscellaneous facilities (container washing facilities, bunkering services, fuel facilities for cargo handling equipment, rest houses for labourers, water supply, water drainage, power supply, reefer container yard, liquid and solid waste collection, etc.).

Industrial environmental services contribute to the value added. The two kinds of industrial services in a modern port are:

- Ship/vehicle related industrial/technical services, such as ship repairing and other technical services;
- Cargo related industrial services, such as export processing zones in or near the port area with attractive commercial conditions, to generate more cargo throughput and more value added for the port (Japan International Cooperation Agency, 1996).

Since ships, cargoes and industrial activities in port areas have commonly been sources of pollution, ports should also be equipped with the necessary facilities for environmental protection.

Administrative/Commercial services should serve the principle that speed and high volume cargo movements require not only that the port be efficient in its management, but also in its procedures, administrative regulations and services. Port administrative efficiency can be evaluated in two groups: (1) Documentation and regulation, (2) Working schedule.

Modern ports, which are used as transhipment ports, may provide free zone status along with efficient administrative services. Export manufacturing/processing activities should be included in the free zone, which can greatly increase administration services both for importing raw materials and exporting finished products.

Commercial services provided include banks, insurance companies, legal services and communication services within the port area. In Rotterdam and Antwerp, some business parks have been created in which there are trade and distribution centres for individual foreign countries (Japan International Cooperation Agency, 1996).

Logistic/Distribution services are incorporated into the conventional, industrial, environmental, administrative and commercial activities in container ports. One new type and typically logistic activity of a modern port today is its distribution service. Cargo and information are two inseparable elements and both need to be distributed (Frankel, 1987). In some modern ports, the distribution centres are called 'distriparks'. This concept involves bringing together companies that specialise in grouping, storage, freight forwarding, international transport, customs documentation etc. onto a site close to a port, road, rail or air transport hub (Japan International Cooperation Agency, 1996).

Modern ports only facilitate access to distribution services and then leave distribution activity to specialised firms. These services facilitated by the port are:

- Warehousing services;
- Water-air-land transport connections;
- EDI (Electronic Data Interchange) services;
- Value-adding activities and services;
- Simplified customs services.

In a modern port, added value can take various forms such as cargo consolidation and deconsolidation, stuffing/unstuffing containers, crating, palletising, shrink-wrapping, labelling, weighing, repackaging etc. For example, at the Port of Singapore in 1990, the value added activities generated about US$ 63 million or 12% of the port's total revenue in 1990 (Japan International Cooperation Agency, 1996).

Meanwhile, port services provided to users, especially to container lines, require a high level of physical facilities and services provided to the lines are interrelated with the physical infrastructure.

Service Quality Determinants

Port services in international container transportation meet the demands of import/export activities, and the service quality produced by the ports for customers, namely the shippers and shipowners, has to meet certain standards for the transportation service to create a competitive advantage. According to the classification of services process (Lovelock, 1996), port services like shipping services can be considered as lying within the 'possession processing services category' in which customers ask a service organisation to provide treatment not for themselves but rather for some physical possession, e.g. in this case, ships or cargoes owned. Container ports serve different types of customers both from local and international markets. They operate within such industrial markets as international trade, transportation and shipping.

The word *quality* has different meanings to people according to the context in which it is used. Garvin identifies five perspectives (Garvin, 1988). The transcendent view of quality, the product based approach, user based definitions, the manufacturing approach, and a value based approach. It is suggested that these alternative views of quality help to explain the conflicts that sometimes arise between managers in different functional departments.

Because of the intangible, multifaceted nature of most services, it may be harder to evaluate the quality of a service than that of a good. Since customers are often involved in service production, the process of service delivery called *functional quality* and the actual output of the service called *technical quality* need to be distinguished (Grönroos, 1990).

The most extensive research into service quality is strongly user oriented. Determinants of service quality have been analysed by those in the marketing discipline and consequently the following ten groups of variables have been agreed (Parasuraman *et al*, 1985; Zeithaml *et al*, 1990; Zeithaml *et al*, 1996): reliability, responsiveness, competence, access, courtesy, communication, credibility, security, understanding/knowing the customer, and tangibles. With the results of further research on service quality these ten dimensions have been combined and reduced to five dimensions namely reliability, responsiveness, tangibles, assurance (combination of competence, courtesy, credibility and security) and empathy (combination of access, communication, knowing and understanding the customer) (Zeithmal *et al*, 1990). These determinants may briefly be grouped as tangibles and intangibles and the importance of these differs according to the nature of the industry. Tangibles play a major role in some industries, whereas intangibles are more important in others.

In ports and especially container terminals, tangibles, namely the physical facilities provided by the port, are important dimensions of service quality. Quality of port operations is also related to the physical infrastructure. The analysis of applications of service quality criteria in port services for container transport will present the port with the data necessary for better quality service.

Level of customer contact is another important factor in service businesses. Services are usually divided into three levels of contact, namely high, medium and low contact services, reflecting actual customer behaviour as it relates to the core product (Lovelock, 1996). Ports may be considered in the range of medium-contact services in which a medium level of contact exists between the port authority and the customers, namely shippers and/or shipowners.

Background

Customer orientation and service quality have been a major area of interest in the port industry in the 1980s. Frankel (1987), Chadwin *et al.* (1990) and the Japan International Cooperation Agency (1996) have each shown that port operation performance and service quality are interrelated to each other.

A survey of the Atlantic market related to the criteria for port choice among shippers and forwarders, indicated that the efficiency of facilities has been of considerable significance in determining a port's competitive position (Peters, 1989). The results of the survey indicate that container flows do not depend solely on the characteristics of the ports directly and additionally costs and service differentials affect port choice decisions. Another analysis of port selection factors stemmed from a survey amongst purchasing managers and showed that they place great emphasis on a port's ability to provide customers with value-added services beyond mere physical operations (Murphy and Daley, 1994). A recent study on market research for container ports indicated that service quality is an important decision parameter for both container lines and shippers, and that the decision power generated by establishing a port hub for a shipping line also influences the choice of a port (Teurelincx, 1998b).

Another research study completed with shippers involved in the export business suggested that the services of maritime agencies are the most important attribute in terms of the quality of port services (Cerit, 2000).

A survey of ports' perceptions of the future conducted in 1991 by the International Association of Ports and Harbors shows that ports regard their relation to their municipal authorities as most important. Road access to ports ranks very highly amongst the concerns of ports as well (Japan International Cooperation Agency, 1996).

Objectives

Previous research on port services has concentrated upon:

- the effects of port services on port performance (Frankel, 1987; Chadwin *et al*, 1990; Japan International Cooperation Agency, 1994; Teurelincx, 1998);
- the effects of port services on competitive advantage of the maritime transport function (Cerit, 2000).

On the other hand service quality determinants have been tested in several service producing industries other than ports (Zeithmal *et al*, 1990).

Analysis of quality factors in ports, especially at container terminals and from the point of view of container ship lines, constitutes the main objective of this study. Due to the outstanding importance of the physical infrastructure and operations carried out in ports, service quality determinants in these two areas have to be analysed by specific studies of the sector and cannot be derived from studies of other industries.

The objectives of the study are determined by the following statements:

- To search for the service quality determinants affecting the competitive position of the container port from the point of view of container ship lines.
- To search for service quality measures concerning the physical infrastructure of the container port from the point of

view of container ship lines.

- To search for service quality measures concerning the operation of the container port from the point of view of container ship lines.
- To measure the level of importance of each specific dimension for container ship lines.

Methodology

To analyse the factors affecting service quality of a container port from the point of view of container lines it was decided to establish a sample of respondents, which had a strong position in container transport and which were noted international representatives of container ship lines.

The Port of Izmir, operated by the Turkish State Railways, faces the Aegean Sea and is situated at the pivotal point of sea trade between Western Europe and North Africa. The Port has a large agricultural and industrial hinterland and plays a substantial role not only as centre for industry and agricultural trade in the Aegean region but also as a vital function in Turkish exports. Loading/unloading capacity of the port is 1,666,200 tons/year general cargo and 4,898,100 tons/year in containers, totalling 6,564,300 tons/year. The outstanding importance of the container terminal facilities in the port has allowed Izmir to attract 388,172 TEU, 35% of all container handling in Turkey, which amounted to 1,001,692 TEU in 1997 in total (DTO, 1998).

Due to the importance of the Port of Izmir in container transport, the container lines situated in Izmir are also strong local representatives of international container ship lines. To assist in achieving meaningful results, it is proposed to carry out the research using the complete population of container ship liner agents in Izmir, and to end with an analysis of the service quality factors present in the container terminal of the Port of Izmir.

Questionnaire Development

A questionnaire consisting of five different parts was developed. The first part covered 17 open-ended and multiple-choice questions concerned with the company profile.

The second part consisted of 18 statements on a 5-point Likert-scale aiming at determining the opinions of liner agents on the physical infrastructure of the container terminal at the Port of Izmir and particularly the role of service quality of liner agents (Table 3).

The third part consisted of 27 questions (five open-ended questions allocating points amongst sub-factors of service quality, 22 statements on a 5-point Likert-scale) concerning the opinions of liner agents on service quality performance factors present in the container terminal of the Port of Izmir (Table 2). In this part the statements covered in previous research on service quality measurements in several service industries were adapted to container port services (Zeithmal *et al*, 1990).

The fourth part consisted of 22 statements on a 5-point Likert scale (Table 4) designed to analyse the operational variables of container terminal services affecting the service quality of container lines. The final part consisted of personal data for the respondents.

Sample

The sample for the study covered the total population of container ship liner agents in Izmir, which came out to be 21 companies. The questionnaires were taken to the general managers of the companies personally and the replies were again collected personally. Twenty of the companies responded to the survey, giving a return rate of 95.23%. In the main it was general managers, operations managers or marketing managers, who replied to the questionnaires.

Data Analysis Procedures

The objectives of the study have been grouped into three aims and adapted to meet the requirements of the factor analytical procedures selected. Different statistical techniques and procedures are used to analyse different types of questions and statements in the questionnaire.

Data processing was carried out using the SPSS (Statistical Package for the Social Sciences) Program. Open-ended and multiple-choice questions aimed at profile construction were analysed by relative frequencies.

For Likert scaled statements directed towards factor analysis, means for sample sizes and standard deviations were calculated to measure the level of importance of each specific dimension for companies.

Results, Evaluation and Discussion

The research completed among the container lines' agents in Izmir produced two groups of data. The first group consisted of information on company profiles; the second group produced results of the factor analysis on service quality determinants.

Company Profile

Table 1 summarises the company profile of the liner agents in the sample. About 55% of the companies have an employment level in the range of 0-20 personnel. 60% of the companies handle below 10,000 TEU annually. Around 70% of the agents represent 1-3 lines and 60% serve 1-15 ships.

Of the liner agents in Izmir, 80% serve lines operating in the Mediterranean region. The lines serving the North America-Canada, South America, North Europe, Africa and Far East regions use transhipment ports; however container ships operating to the Mediterranean area as a whole, provide direct services. The majority of container ships calling at the Port of Izmir are feeder ships, accounting for 68% of the companies. 47% of the liner agents indicated that they serve mother ships.

Factor Analysis

The objectives of the study covered three factor analyses concerning service quality determinants affecting the competitive position of Izmir container port, service quality measures concerning the physical infrastructure of the port and service quality measures concerning the operation of the port from the point of view of container ship lines.

Table 2 analyses the six sets of factors obtained through the factor analysis of the statements concerning service quality determinants affecting the competitive position of the container port.

The factors extracted explain 87.52% of the variation in the data. Considering the six factor groupings it can be noted that highest loadings are associated with the behaviour of port employees, service at times promised, prompt services of the employees in the port, neat appearance of the employees, visually appealing materials associated with the service and interest in personal requirements. The second most important attribute is the employees' interest in the personal requirements of the liner agents. In terms of the frequencies of the responses given to the Likert-type statements of Table 2 error-free records, employees' promise on when services will be performed, employees' willingness to respond to requests and employees' personal attention are the most important.

Table 1: Company Profile of Container Line Agents in Izmir

Variable			Variable	
Personnel	n	%	*Containers Handled*	n
0-20	11	55.0	0-10,000	12
21-40	5	25.0	10,001-20,000	4
41-60	1	5.0	20,001-30,000	1
61-80	2	10.0	30,001-40,000	2
80+	1	5.0	40,001-50,000	-
Total	20	100	50,001-60,000	1
			Total	20
Lines			Ships Served	
1-3	14	70.0	1-15	12
4-5	2	10.0	16-30	3
6-7	2	10.0	50 +	1
8-9	1	5.0	Missing	4
10-15	1	5.0	Total*	20
Total	20	100		
Lines Served **			Transhipped Services	
North America	13	65.0	N. America and Canada	10
South America	7	35.0	South America	7
North Europe	12	60.0	North Europe	5
Mediterranean	16	80.0	Mediterranean	3
Russia	10	50.0	Russia	6
Gulf	10	50.0	Gulf	9
Africa	8	40.0	Africa	7
Far East	11	55.0	Far East	11

* Missing cases are not included in the relative frequencies
**Relative frequencies are based on total responses

When compared with the results obtained in previous research on service quality performance measurements (Zeithmal *et al*, 1990), the results in Table 2 differ mainly in two factors, namely tangibles and empathy. The present study suggests that amongst the service quality factors present in the container port services, factors related to infrastructure/tangibles and empathy fall into other factor groups. This fact is consistent with the importance of physical infrastructure in port services and the relatively low level of customer contact in container port operations and services.

As a matter of fact, allocation of points among sub-factors of service quality proved that infrastructure was considered as 39% important by respondents, whereas reliability received only 22% importance. Responsiveness was considered as 14% important, with the lowest rank achieved by assurance (12%) and empathy (13%).

Table 3 analyses the factors, and reflects the effects of physical infrastructure at the Port of Izmir, and relates these dimensions to the competitive position of the companies. The factors obtained explain 86.91% of the variation in the data. Seven factor groupings have been obtained where highest loadings are found for covered storage facilities, towage and pilotage facilities, operational quality of cargo handling and storage facilities, information processing facilities, power supply and lighting, road connections and parking areas and draught. When the frequencies of the responses given to the Likert-type statements of Table 3 are considered, road connections is the most important attribute. Operational quality and quantity of cargo handling and storage equipment, information processing facilities and railway connections were the other leading attributes.

Table 4 analyses service quality measures concerning the operation of the container port from the point of view of container ship lines and relates these dimensions to the competitive position of the companies. Four factor groupings have been obtained explaining 84.15% of the variation in the data. Highest loadings were achieved on container clearance, sanitary application, information services and personal safety. With respect to the responses of the liner agents, it was observed that container reception and distribution received the highest frequency of response. The second attribute is found in stuffing/stripping operations, sanitary application and customs permissions. In third place are the container loading/discharging and immigration applications.

Reliability Analysis

To check the internal consistency of the sample and the scales, a reliability analysis was performed for the factor groupings, and the results are given in the tables as alpha values.

The high score of the alpha values for all the cases prove that reliability of the sample is quite high. In Table 3 the factors concerning cargo handling and storage equipment and road connection and parking areas have received a low alpha level, and this has affected the overall alpha of the factor grouping. Factor groupings in Table 2 and Table 4 have received high alpha levels.

Discussion

Three main groups of factor analysis have been completed. These have shown that the service quality of liner agents is affected by the quality of container port services and this study has presented the 17 factors affecting the whole system. The total of 62 statements initially put forward, has been reduced to only 17 factors, and moreover the frequencies of the responses relating these factors have also been obtained.

With developments in containerisation, investment in container ports has been the major area of interest in maritime transportation and due to the outstanding rates of investment, it is of vital importance for the port authorities to carry out their services in a differentiated way to maintain their competitive position. Conditions of severe price competition in the container markets have forced service suppliers to improve their services with respect to developed quality. Recent studies on port performance measures have proved that the quality of port services remains at the highest level of port selection criteria for the shippers.

The results of the study have shown that container port services, which occupy a medium level of contact with customers, depend largely upon physical infrastructure of the port and thus differ from service industries where a high level of contact with customers take place. They differ from those services where tangible infrastructure is of less importance.

The factor analyses in this study produced seven factor groupings relating physical infrastructure and container ship service quality. Among the main facilities of the port, storage facilities, operational quality of cargo handling and storage equipment, information-processing facilities, and road and railway connections received the highest loadings and highest frequencies. These factors all result in speedier and better scheduling of container services, any delay ending in unavoidable costs and decline in market shares.

The analysis of the factors examining the effects of port operations on container ship liner service quality has ended with four factor groupings amongst which container clearance, reception and distribution besides the stuffing/stripping and loading/discharging operations, have received the highest loadings. The highest frequencies which were obtained in sanitary applications, customs permissions and immigration applications, indicate areas to be improved.

Limitations and Further Study

This study approached the analysis of container port service quality in terms of the competitive position of liner agents. As an introductory study to analyse the views of liner agents as representatives of lines, the study provided a general analysis of the determinants in the area. Further studies could be accomplished on each of these determinants in depth and related service quality measures might be obtained.

Besides liner agents, a comparative study among shippers and other users of the port would help to provide an integrated view of port services. Carrying out the same study in a port of higher container throughput where a population of larger liner agents exists, would help to produce more statistically significant results. To test conditions in general cargo terminals, a corresponding study could also be accomplished.

Keeping in mind that the study was carried out on the services of a government port, certain differentiated results may be obtained if a similar study examined the services of privately or autonomously operated ports.

Container transportation retains a leading role in world seaborne trade and thus container ports, as the nodal points of the logistical chain, need to be evaluated within a service quality perspective.

Table 2: Factor Analysis on the Service Quality Determinants of Container Ship Services at the Port of Izmir

Factor Groupings	α	Mean*	S.D.	Factor Loadings					
				I	II	III	IV	V	VI
Service Quality Factors	0.9121								
I Assurance and Empathy	0.8732								
Behaviour of port employees		2.6154	0.9608	0.8970					
Individual attention		2.6923	1.1821	0.8640					
Courteousness of port employees		2.6923	1.1094	0.8520					
Safe transactions		2.6154	1.1929	0.8100					
Error-free records		3.3846	1.1929	0.6190					
Our best interest at heart		2.6923	1.1821	0.5900					
Convenient operating hours		2.4615	0.9674	0.5850					
II Reliability	0.9036								
Services at times promised		2.9231	1.1152		0.9480				
Sincere interest in solving problems		2.9231	1.1152		0.8370				
Performing the service right the first time		2.6923	1.0316		0.7910				
Employees have knowledge to answer questions		3.3077	1.1094		0.7680				
Doing something at the time it is promised		2.6154	1.1209		0.7470				
Modern looking equipment		2.3077	1.1821		0.5090				

Table 2 Continued

	α	Mean*	SD	I	II	III	IV	V	VI
III Responsiveness	0.8284								
Employees in port give prompt service		2.7692	0.9268			0.8400			
Employees in port are always willing to help		2.8462	0.9871			0.8020			
Employees tell us exactly when services will be performed		3.3846	1.1209			0.6950			
Physical facilities are visually appealing		2.3077	1.1094			0.6770			
IV Port Labour	0.7020								
Employees have neat appearance		2.2308	1.3009				0.8270		
Employees are never too busy to respond to requests		3.3846	1.1209				0.7590		
V Attention	0.7524								
Materials associated with the service are visually appealing		2.7692	0.9268					0.9510	
Employees pay personal attention		3.3846	1.1929					0.7990	
VI Personal Requirements	-	3.6154	0.9608						0.5720

*5-point Likert scale- 1: Completely Disagree, 5 : Completely Agree

Table 3: Factor Analysis on the Effects of Physical Infrastructure of the Port of Izmir on Containership Liner Service Quality

Factor Groupings	α	Mean*	S.D.	Factor Loadings						
				I	II	III	IV	V	VI	VII
Physical Infrastructure Factors Affecting Liner Service Quality	0.8158									
I Container Berth and Storage Area	0.8770									
Covered Storage Facilities		3.3333	1.1376	0.9610						
Open Storage Facilities		3.3333	1.2834	0.9310						
Specialised Container Storage Facilities		3.7222	1.2744	0.7800						
Berth Characteristics		3.5556	1.3382	0.7730						
II Port Auxiliary Services	0.8665									
Towage and Pilotage		3.1111	1.4096		0.9210					
Water, Bunkering		2.8333	1.2948		08510					
Communication Facilities		3.1667	1.4653		0.7320					
Container Repair and Maintenance		3.3333	1.2367		0.4560					

Table 3 Continued

	α	Mean*	SD	I	II	III	IV	V	VI	VII
III Cargo Handling and Storage Equipment	0.3530									
Operational Quality		4.5000	07071			0.7590				
CFS Quality		3.7222	1.2274			0.7560				
Quantity		4.5000	1.0432			0.4740				
IV Limited Services	0.5268									
Information Processing Facilities		4.3333	1.0847				0.7560			
Railway Connection		4.1667	0.8575				0.7400			
Container Cleaning Facilities		3.5000	1.2948				0.6040			
V Power Supply and Lighting		3.3333	1.1376					0.9330		
VI Road Connection and Parking Areas	0.4439									
Parking Areas		3.8889	1.2314						0.8860	

*5-point Likert scale- 1: Completely Disagree, 5: Completely Agree

Table 4: Factor Analysis on the Effects of Port Operations on Containership Liner Service Quality

Factor Groupings	α	Mean*	SD	Factor Loadings			
				I	II	III	IV
Factors of Port Operations Affecting Liner Service Quality	0.9565						
I Call for Port Service and Scheduling of Terminal Processes	0.9457						
Container Clearance		3.9231	1.4412	0.9370			
Call for Port Service		3.6923	1.4936	0.9190			
Call for Towage and Pilotage		3.6154	1.5566	0.8350			
Container Loading/Discharging Operations		4.0000	1.2910	0.7930			
Immigration and Custom Control		3.8462	1.4632	0.7810			
Container Reception and Distribution		4.1538	1.4632	0.7760			
Container Stowing and Storage		3.6923	1.1821	0.6920			
Container Transfer		3.6923	1.3775	0.6800			
Towage and Pilotage Operations		3.9231	1.5525	0.6480			
Stuffing/Stripping Operations		4.0769	1.3205	0.5910			

Table 4 Continued

	α	Mean*	SD	I	II	III	IV
II Clearance Schedule	0.9230						
Sanitary Application		4.0769	0.8623		0.8780		
Seaworthiness Obtain		3.8462	0.8987		0.8290		
Immigration Application		4.0000	1.0000		0.7920		
Customs Permissions		4.0769	1.1875		0.7740		
Sanitary Permissions		3.8462	1.4051		0.7360		
Berthing Permission		3.5385	1.3301		0.6940		
Customs Officers' Permission		3.8462	1.4632		0.6240		
III Information Flow	0.8663						
Information Services		3.5385	1.2659			0.9300	
Information Flow		3.9231	1.2558			0.8010	
IV Security and Safety	0.7706						
Safety		3.2308	1.0127				0.8640
Security Against Theft		3.0769	1.0377				0.7260
Deposits for Services		3.8462	1.2810				0.6350

*5-point Likert scale – 1: Completely Disagree, 5: Completely Agree.

References

Bucur, I. (1998) *Cities and Port Competitiveness Actions with Impact on the Performance of the Port Operations and Management*, Technonav'98, Constanta: Ovidius University, Volume I, 338-347.

Cerit, A. G. (2000) Maritime Transport as an Area of Competitive Advantage in International Marketing, *International Journal of Maritime Economics*, 1, 3, 49-67.

Chadwin, M.L., Pope, J.A. and Talley, W.K. (1990) *Ocean Container Transportation: an Operational Perspective*, New York: Taylor and Francis Inc.

DTO (1998) *Deniz Sektörü Raporu 1997*, Istanbul: DTO Publication.

Evangelista, P. and Morvillo, A. (1998) Logistical Integration and Co-operative Strategies in Liner Shipping: Some Empirical Evidence, *8th World Conference on Transport Research* 12-17 July, Antwerp.

Frankel, E.G. (1987) *Port Planning and Development*, New York: John Wiley and Sons Inc.

Garvin, D.A. (1988) *Managing Quality*, New York: The Free Press.

Grönroos, C. (1990) *Service Management and Marketing*, Lexington: Lexington Books.

Japan International Cooperation Agency (1994) For The Third Generation Ports, *Port and Harbor Engineering II.*

Japan International Cooperation Agency (1996) Containerization and Liner Shipping, Global Alliance and Port Management, *Seminar on Port Administration and Management.*

King, J. (1997) Globalization of Logistics Management: Present Status and Prospects, *Maritime Policy and Management*, 24, 4, 381-387.

Lovelock, C. (1996) *Services Marketing*, London: Prentice Hall International Editions.

Marchese, U. (1997) Intermodality and the Evolution of Competition in Shipping Markets, *International Conference On Ship and Marine Research, NAV' 97, University of Naples Federico II*, Sorrento, 1.3-1.14.

Marchese, U., Musso, E. and Ferrari, C. (1998) The Role for Ports in Intermodal Transports and Global Competition: A Survey of Italian Container Terminals, *8th World Conference on Transport Research*, 12-17 July, Antwerp.

Mester, B. (1991) Marketing from the Port's Point of View, *Port Management Textbook, Vol.3*, Bremen: ISL.

Murphy, P.R. and Daley, J.M. (1994) A Comparative Analysis of Port Selection Factors, *Transportation Journal*, Fall, 15-21.

Notteboom, T. and Winkelmans, W. (1998) Spatial (De)Concentration of Container Flows: The Development of Load Center Ports and Inland Hubs in Europe, *8th World Conference on Transport Research*, 12-17 July, Antwerp.

OCDI-The Overseas Coastal Area Development Institute of Japan (1990) Container Terminal Planning, *Port Planning and Development*.

OCDI-The Overseas Coastal Area Development Institute of Japan (1996) Container Terminal Planning, *Seminar on Port Administration and Management*.

Okada, H. (1990) *Port Planning and Development*, Tokyo: OCDI (Overseas Coastal Area Development Institute of Japan).

Parasuraman, A.B., Leonard, L. and Zeithaml, V.A. (1985) A Conceptual Model of Service Quality and its Implications for Future Research, *Journal of Marketing*, 49, 41-50.

Peters, H.J. (1989) *Seatrade, Logistics, and Transport*, Washington D.C: The World Bank Policy and Research Series.

Slack, B., Comtois, C. and Sletmo, G. (1996) Shipping Lines as Agents of Change in the Port Industry, *Maritime Policy and Management*, 23, 3.

Song, D. and Cullinane, K. (1998) Port Ownership and Productive Efficiency: The Case of Korean Container Terminals, *8th World Conference on Transport Research*, 12-17 July, Antwerp.

Tamvakis, M.N. and Thanopoulou, H.A. (1998) Does Quality Pay? The Case of the Dry Bulk Market, *8th World Conference on Transport Research*, 12-17 July, Antwerp.

Teurelincx, D. (1998a) Functional Analysis of Port Performance as a Strategic Tool for Strengthening a Port's Competitive and Economic Potential, *8th World Conference on Transport Research*, 12-17 July, Antwerp.

Teurelincx, D. (1998b) The Use of Market Research and Simulation Techniques in Port Capacity Planning, *8th World Conference on Transport Research*, 12- 17 July, Antwerp.

UNCTAD (1995) *Marketing Promotion Tools for Ports*, New York: United Nations Publication.

Uyguç, N. (1998) *Hizmet Sektöründe Kalite Yönetimi*, Izmir: Dokuz Eylül Yayinlari.

Zeithaml, V.A., Berry, L.L. and Parasuraman, A. (1996) The Behavioral Consequences of Service Quality, *Journal of Marketing*, 60, 31-46.

Zeithaml, V.A., Parasuraman, A. and Berry, L.L. (1990) *Delivering Quality Service*, New York: The Free Press.

13 Comparative Analysis of Recently Privatised Major Ports in Turkey

FUNDA YERCAN AND KAZIM YENI
SCHOOL OF MARITIME BUSINESS AND MANAGEMENT
DOKUZ EYLUL UNIVERSITY, IZMIR

Introduction

Major ports in Turkey are owned and operated by two state institutions, Turkish State Railways and the Turkish Maritime Organisation and regularly suffer from heavy financial losses. As a consequence, the idea of privatisation within the maritime sector in Turkey has been widely adopted in the last decade. Consequently, this study reviews the recently privatised ports in Turkey in an attempt to highlight their development and the importance of privatisation within the industry.

The majority of the institutions, organisations and companies involved in the maritime sector in Turkey, and vessels in the merchant fleet, are owned and operated by organisations in the private sector. The management system of these companies and the methods of business operation are based upon the principles arising from a highly competitive free market situation. Therefore, the shipping industry benefits from the dynamism of the entrepreneurial character of the private sector. On the contrary, the major ports are owned and operated by state institutions which we have seen, tend to suffer from heavy losses each year.

The major ports in Turkey, including their general specifications and indicators of total trade undertaken, are reviewed briefly in the following section. In addition, some brief explanation is put forward for the major state-owned and operated ports, major private ports and recently privatised ports. Major laws and regulations related to privatisation actions in Turkey are reviewed in addition. Recent privatisation of the 11 ports previously owned by the Turkish Maritime Organisation, is analysed in a later section of this study. Subsequently, the advantages and disadvantages of

208

privatisation within these ports are discussed.

The Major Ports in Turkey

Ports in Turkey consist of 15 major government ports, 50 small piers under the control of municipalities and 60 private ports and piers. Recent total trade indicators including both foreign trades and cabotage shipping activities to and from all of these ports in Turkey are illustrated in Table 1. Approximately, 22% of total trades was handled as export cargo, 52% as import cargo and 26% as cabotage cargo in 1999 (Chamber of Shipping, 2000). The majority of these cargoes were handled at major ports, which are grouped under state controlled ports.

Table 1: Total Trades Undertaken at Ports in Turkey (million tons)

	1994	1995	1996	1997	1998	1999
Imports	48.805	58.502	65.290	70.140	79.303	77.978
Exports	20.932	19.760	18.035	20.415	24.773	32.923
Cabotage	31.905	32.709	34.208	38.215	38.715	38.658
Total	101.641	110.971	117.533	128.770	412.791	149.559

Source. Chamber of Shipping (2000).

Major State-owned Ports

The major ports under the control of the government are either owned and operated by Turkish State Railways (TSR) or by the Turkish Maritime Organisation (TMO). The seven general purpose major ports, which are owned and operated by Turkish State Railways and thus indirectly under the control of the Ministry of Transport, are the Ports of Istanbul (also known as the Port of Haydarpasa), Derince and Bandirma on the Sea of Marmara to the northwest of the country, the Port of Izmir on the Aegean Sea to the west, the Ports of Mersin and Iskenderun on the Mediterranean Sea to the south and the Port of Samsun on the Black Sea to the north of the country.

Approximately 90% of the total cargo handled at state-owned ports in Turkey is handled at the Ports of the Turkish State Railways. All of these ports have links with land and railway transportation routes and some are even positioned close to airports, such as the Ports of Istanbul and Izmir, which make them ideally placed from a logistical point of view. In addition, all these state-owned ports have been traditional crossing points for trading and cultural activities between the two continents of Europe and Asia for centuries, the Port of Izmir since ancient times, in particular. As a consequence, these ports act as the major gateways for the country to the world whilst playing a key role in the international and domestic transportation systems and foreign trade of Turkey.

Ports of the Turkish State Railways have a 30 million ton capacity for cargo handling with a 34 million ton annual capacity of general cargo storage and a container storage capacity of 1,080,000 TEUs/year (Yercan and Roe, 1999a). Around 11,600 ships can be accommodated annually by these ports with a total berth capacity of 50 million tons based upon a continuous three-shift operation system. Approximately 28 million tons of cargo are handled annually at these ports with the majority of cargo types being dry bulk, general cargo and containers. The general specifications and total capacities of these ports are summarised in Table 2. In addition, the total cargo handled at these ports in 1999 is illustrated in Table 3.

Table 2: General Specifications and Capacities of the Ports of Turkish State Railways

Ports	Berth length (m)	Max. drt (m)	Total ship call capacity		Cargo handling capacity		
			Cargo	Pass.	Dry bulk/ general (000 tons/pa)	Cont-ainers (000 tons/pa)	Cont-ainers (TEU/ pa)
Haydarpasa	2,765	-12	2,651	-	2,834	3,082	354,000
Derince	1,132	-5	1,105	-	1,799	-	40,000
Samsun	1,756	-12	1,130	-	2,189	-	40,000
Mersin	3,180	-14.5	2,650	623	2,639	2,855	266,000
Iskenderun	1,427	-12	640	-	3,224	-	20,000
Bandırma	2,788	-12	1,037	3,240	2,636	-	40,000
Izmir	2,959	-12	2,389	1,246	1,469	4,082	443,000
Total	16,007	-	11,602	5,109	16,790	10,019	1,203,000

Sources: Chamber of Shipping (2000); Turkish State Railways (2000).

Besides the ports of Turkish State Railways, the remaining eight government ports, which are considered secondary major ports, were actually owned, managed and operated by the Turkish Maritime Organisation and are situated as follows: the Port of Tekirdag on the Sea of Marmara to the northwest, the Ports of Sinop, Ordu, Giresun, Trabzon,

Rize and Hopa on the Black Sea to the north and the Port of Antalya on the Mediterranean Sea to the south of the country. In addition, there are also six other ports within this organisation, which are considerably less significant compared with the others: the Ports of Dikili, Cesme, Kusadasi, Gulluk and Marmaris on the Aegean Sea and the Port of Alanya on the Mediterranean Sea. Amongst these ports, the Ports of Hopa, Rize, Giresun, Ordu, Sinop, Tekirdag, Antalya, Kusadasi, Cesme, Marmaris and Dikili were recently privatised and will be considered further in the following section.

Table 3: Activities of the Ports of Turkish State Railways (1999)

Ports	Total ship calls	Total cargo handling		
		Dry bulk and general cargo (000 tons/year)	Container (000 tons/year)	Container (TEUs/year)
Haydarpasa	1,977	2,975	2,713	277,233
Derince	404	661	38	5,501
Samsun	985	2,055	10	1,904
Mersin	3,907	4,325	2,756	251,188
Iskenderun	489	1,367	2	379
Bandırma	2,147	3,096	-	-
Izmir	2,725	2,649	4,245	435,962
Total	12,634	17,128	9,765	972,167

Source: Turkish State Railways (2000).

The state-owned and operated ports offer a full range of services which includes amongst other facilities, pilotage and towage, quay occupation, fresh water supply, solid and liquid waste removal, handling services, storage, commodity weighing and the rental of equipment. Stevedoring services are mostly supplied by the port authority and the existence of a private stevedoring service is not allowed at most of the state ports. Port tariffs are calculated by the Turkish Maritime Organisation and port charges may slightly differ from one port to another. A vessel calling at a port in Turkey is charged for pilotage (both entering and leaving the port), tugging (in/out of port), quay dues, waste removal, sanitary dues, light dues (again both entering and leaving the port), a chamber of shipping fee, an agency fee, an attendance/supervision fee and a number of other charges depending upon the types of cargoes loaded and unloaded, i.e. transshipment fee, a commission on inward freight, freight tax, forwarding fees, harbour dues etc. (Yercan and Roe, 1999a).

Ports under the control of municipalities tend to be comparatively small and are limited to a small volume of coastal traffic serving the local needs of coastal cities and towns. The majority of these ports are situated either to the north or northwest with 20 of them on the Black Sea and 17 of them on the Sea of Marmara with the remaining seven on the Aegean Sea and six on the Mediterranean Sea.

Major Private Ports

The major private ports amongst the 60 non-state owned ports in Turkey are mostly situated on the Sea of Marmara to the northwest of the country. The most significant ones can be identified as the Ports of Gemport, Armaport, Kumport, Akport, Arkas, Anatolian Cement Port, Mardas, Sedef, Colakoglu, Poliport and Bagfas in addition to a number of other, but less significant private ports (Yercan and Roe, 1999b).

The Port of Gemport was established in 1992 as the first private port in Turkey and is under the operation of Gemlik Port and Warehousing Administration Company. Cargo handling capacity is 2,600,000 tons/year with an 85% of capacity utilisation. Total cargo storage areas are 140,000 m^2 open and 2,000 m^2 closed areas with 5,000 TEU container stock capacity. The port concentrates upon general cargo with container terminals located in the Gulf of Gemlik and is connected to its hinterland by a good highway network. Berthing for ro-ro vessels is also available through a combined terminal for car carriers.

The Ambarli Port Complex - or the Ports of Ambarli, consists of the Ports of Kumport, Armaport, Anatolian Cement Port, Soyakport and Mardas. The Port of Ambarli is located on an area of 1.5 million square metres in Ambarli in the Marmara region. The substructure, superstructure and highway connections at these private ports were constructed by their private owners and operators in general with a total investment of US$400 million (Chamber of Shipping, 1998) without any subsidies or incentives provided by the government. These ports have been established as the major competitors of and the alternatives to the Port of Haydarpasa/Istanbul stemming from the location of the Port of Haydarpasa/Istanbul at the Bosphorus where heavy traffic takes place.

The Port of Armaport of the Ambarli Port Complex was established in 1997 through the co-operation of shipowners with the aim of supplying the rapidly increasing demand for port services. The operator of the ports is known as the Shipowners Port Management Company. The port supplies the increasing needs of a range of dynamic services in the Marmara region with increasingly competitive prices and modern technological equipment in competition with the much higher port tariffs and excessive container traffic that characterises the Port of Haydarpasa/Istanbul (The World, 1997). A major competitor for the port in international markets is the Port of Piraeus in Greece, which has developed into a hub port for the Russian Federation and other CIS countries.

The Port of Kumport is situated to the north of the Sea of Marmara with its own multipurpose terminal. The port is operated by the Association of Sea Sanders in Istanbul. The container terminal of the port is operated by a group formed by Kumport Port Management and Arkas Shipping Group - one of the leading Izmir based group of shipping companies dealing in freight forwarding, logistics, shipowning and ship operating sectors of the shipping industry (Cargo News, 1998).

Recently Privatised Ports

Privatisation of the major ports is expected to be centred on the privatisation of secondary major ports encouraging the major ones to follow. The ports already privatised between 1997 and 2000 are the secondary major ports - the Port of Tekirdag on the Sea of Marmara, the Ports of Sinop, Ordu, Giresun, Rize and Hopa on the Black Sea, the Port of Antalya on the Mediterranean Sea and the Ports of Kusadasi, Cesme, Marmaris and Dikili on the Aegean Sea. All of these ports were previously owned and operated by the Turkish Maritime Organisation and are currently leased to and operated by a number of companies, i.e. Akport-Tekirdag Port Organisation, the Port Organisations of Sinop, Ordu and Giresun established by Cakiroglu Building and Construction Company, Riport-Rize Port Organisation and Investment established by three shipping groups of companies - Yardimci, Ergencler and Cillioglu Groups, Park Shipping and Hopa Port Organisation, Middle East-Antalya Port Organisation, Limas Port Company, Cesme Port and Tourism Operations, Marmaris Common Entrepreneurship Group and Dikili Port and Tourism Operations (Chamber of Shipping, 2000).

In addition, the Ports of Trabzon, Alanya and Gulluk were scheduled to be privatised in 2000; however, some delay in privatisation actions will take place due to a delay in the confirmation of the procedure by the Directorate of Privatisation Administration.

Privatisation of Ports

Similar to many governments in today's world, privatisation has been a key policy element and was adopted in Turkey after rapid progress in many industries and economic liberalisation policies after the 1980s. There has been development in government policy to privatise many state institutions which previously suffered from heavy losses, causing problems for the economy. The government adopted a privatisation policy as a major tool to reduce the role of the public sector. Main considerations have been the need for improved services and projects to be managed better by the private sector.

The privatisation action of state-owned institutions is largely based on law no. 4046 dated 24.11.1994 and law no. 4105 dated 02.05.1995. The major objective of these laws is to privatise the institutions thus increasing productivity, efficient use of resources and current capacity, increasing competitiveness in international and domestic markets, increasing the contribution to the regional and national economy and reducing the continuous increase in the costs and heavy administrative loads of state institutions (Demirbas, 2000).

The procedures of port privatisation have been handled under the control of the Directorate of Privatisation Administration. Privatisation policy for these state institutions - the Turkish Maritime Organisation in this case - focuses in general on the idea of a privatisation programme based on a European model rather than other models such as the UK system (Lloyd's List, 1999). The policy for the privatisation programme adopted in Turkey mainly concentrates upon the idea of leasing the assets of a port to a private company under a long-term agreement where land and basic infrastructure are owned by the state institution. The privatisation action covers the leasing of the operation for 30 years and fully transferring the operation to a private company after the completion of this period; however, this does not involve transferring the ownership of land (The Working Group on Coastal Facilities and Ports, 2000). In addition, cargo handling and other types of equipment and vehicles are also covered by the privatisation action. In the meantime, the private company is responsible for the management of the port including cargo handling, personnel and all additional investments in cargo handling equipment and warehousing (Yercan, 1999).

Analysis of Recently Privatised Ports

Privatisation of the seven ports of the Turkish Maritime Organisation was completed in 1998 with most of them being situated on the coast of the Black Sea. The privatisation of the remaining four ports previously owned by the same institution and situated on the Aegean Sea was completed in 2000. The privatisation action of these ports covers the leasing of operation of ports by private companies for 30 years, as noted earlier. The new operators of the ports, which were formerly owned and operated by the Turkish Maritime Organisation, are listed in Table 4. In addition, physical characteristics and capacities of these ports are illustrated in Table 5.

Table 4: New Operators of Recently Privatised Ports

Port	Operator	Date of privatisation	Period of lease for operation (years)
Hopa	Park Shipping and Hopa Port Organisation	17.06.1997	30
Giresun	Cakiroglu-Giresun Port Organisation	30.06.1997	30
Ordu	Cakiroglu-Ordu Port Organisation	30.06.1997	30
Sinop	Cakiroglu-Sinop Port Organisation	30.06.1997	30
Rize	Riport-Rize Port Organisation and Investment	07.04.1997	30
Tekirdag	Akport-Tekirdag Port Organisation	17.06.1997	30
Antalya	Middle East-Antalya Port Organisation	31.08.1998	30
Kusadasi	Limas Port Company	12.07.2000	30
Cesme	Cesme Port and Tourism Operations	12.07.2000	30
Marmaris	Marmaris Common Entrepreneurship Group	12.07.2000	30
Dikili	Dikili Port and Tourism Operations	12.07.2000	30

Sources: Chamber of Shipping, (2000); JICA and SPAC, (2000).

Despite the fact that the substructure, superstructure and highway connections at private ports are normally constructed by their private owners and operators, there exists an exception in the recently privatised ports. The structures of these ports were already provided by the state because they were previously owned and controlled by the Turkish Maritime Organisation. However, some new facilities and cargo handling equipment have been provided by the new operators after the privatisation of the operation of these ports by a leasing method for 30 years, as noted earlier. A number of specifications of these recently privatised ports are given below.

Table 5: General Specifications of Recently Privatised Ports

Port	Berth length (m)	Max. draught (m)	Cargo handling capacity (tons/year)	Ship call capacity
Hopa	1,145	-10	1,394	1,425
Giresun	1,022	-10	1,394	1,575
Ordu	269	-9	865	350
Rize	130	-5	529	140
Sinop	197	-12	--	250
Tekirdag	1,014	-9	2,900	1,050
Antalya	1,900	-10	3,338	2,975
Kusadasi	920	-10	--	1,741
Cesme	480	-10	--	1,060
Marmaris	462	-12	--	1,460
Dikili	168	-8	193	175

Sources · Chamber of Shipping (2000); Chamber of Shipping (1999).

Improvements in the administrative structure: A professional mentality is adopted in marketing and management of ports resulting from skilled and experienced personnel; wages and salaries of personnel are based on qualifications; training of workers is carried out effectively in the use of sophisticated equipment; modernised and advanced computer systems and high technology communication equipment are used; regular port traffic policies and price tariffs are implemented; the number of personnel was considerably decreased to improve efficiency (Yercan, 1999).

Improvements in service facilities: Loading/unloading of ships takes much less time compared with the period prior to privatisation; sufficient standardisation of cargo handling equipment; improved and sufficient storage facilities for special cargoes; investment for container ships; planning and establishing new projects for better services; modernising facilities at ports (Yercan and Roe, 1999a).

Privatisation of secondary major ports - previously owned and operated by the Turkish Maritime Organisation - was expected to encourage and give rise to the privatisation of the major ports in Turkey, which are currently owned and operated by Turkish State Railways, as noted earlier.

Total volume of cargo handled at recently privatised ports between 1994 and 1999, which covers the total volumes during the periods before and after privatisation, is illustrated in Table 6. It should be noted that the figures for the Ports of Kusadasi, Cesme, Marmaris and Dikili are not given in the table as they specialise in passenger shipping.

Table 6: Total Volume of Cargo Handled at Recently Privatised Ports 1994-1999 (tons)

Ports	1994	1995	1996	1997	1998	1999
Hopa	326,997	311,000	387,000	390,000	--	145,876
Giresun	123,448	122,000	185,000	186,000	--	164,937
Ordu	49,774	46,000	104,000	157,000	--	219,507
Rize	174,918	251,000	298,000	361,000	--	228,151
Sinop	5,728	8,000	7,000	8,400	--	15,905
Tekirdag	1,519,839	1,407,000	1,698,000	1,872,000	--	2,139,417
Antalya	1,198,868	1,579,000	1,657,000	1,662,000	943,000	757,738
Total	3,399,542	3,724,000	4,336,000	4,636,400	943,000	3,939,699

Sources : Chamber of Shipping (2000); Chamber of Shipping (1999).

It is clear from Table 6 that total volume of cargo handled at the Port of Ordu increased by 40% in 1999 after the privatisation action. The figures at the Port of Sinop, whose operator is the same as that of Port of Ordu - The Cakiroglu Group - doubled in the same year; however, total volume at the Port of Giresun decreased slightly by 10%.

It is also clear from the same table that total volume of cargo handled at the Port of Hopa decreased by more than half in 1999. The major reason of this decline is that the port is situated at the far eastern end of the Black Sea Region on the north coast of the country and is located very close to the border with Georgia, causing difficulties in reaching Turkey's major hinterland. There was a decline of 40% in the same figure at the Port of Rize in 1999 with the major reason for the decline stated as the competition between recently privatised ports in the region, many of which are situated close to each other.

The Port of Tekirdag is one of the major competitors for the Port of Haydarpasa/Istanbul on the Bosphorus and for the industrial private ports on the Sea of Marmara coast to the northwest of the country. The Port of Tekirdag has greatly benefited from its location at the entrance to the Sea of Marmara and the shipping routes which connect with other European ports. This is clear from the increase of 20% in total volume of cargo handled in 1999. The Port of Antalya was privatised in 1998 and therefore, rehabilitation has not yet been completed as can also be seen from Table 6 illustrating a large decline in the total volume of cargo handled in 1999.

Conclusions

The maritime sector has rapidly developed as one of the leading industries in Turkey. This development stems from the improvement of foreign trade after the liberalisation policy in the economy from the beginning of the 1980s. In addition, dynamism within the private sector has also helped the development of the industry. Rapid development within the maritime industry in general has occurred particularly in the shipping industry; however, progress in the ports sector has been limited by contrast either because they are still controlled by state institutions, or because rehabilitation actions at recently privatised ports are not yet complete. These rehabilitation actions concentrate upon improvements in administrative structure and service facilities as noted earlier. It will be necessary to take time in evaluating port privatisation as the actions only started in 1997.

As far as the analysis of this study so far, recent privatisation actions have had a positive impact upon operational activities. Operating revenue has started to increase in port operator companies. The operators of recently privatised ports can make port investment decisions with some flexibility and according to their priority requirements. The supply, maintenance and repair of cargo handling equipment has been given priority, contributing positively to improving the level of cargo handling efficiency (JICA and SPAC, 2000).

On the other hand, insufficient level of productivity, lack of computerised processing and insufficient skilled personnel at most of the recently privatised ports are still the major problems to be solved. In addition, fluctuations in the economy of Turkey have direct influence upon the total foreign trade made within the ports, and which has caused a general decline in the total volume of cargo handled. Furthermore, coordination between state and private ports needs to be improved to prevent overlaps in investment for common hinterlands. Finally, recently privatised ports could also be considered in the nationwide frame of state port development in Turkey as the investment in construction or maintenance of new facilities is currently difficult to bear for private companies until they become more firmly established.

References

Cargo News: Transportation and Logistics (1998) May, Istanbul, Turkey, pp.1-15.

Chamber of Shipping (1998) *Turkish Shipping World*, December, Istanbul, Turkey, pp.12-19, 167-168.

Chamber of Shipping (2000) *Report on Shipping Sector - 1999*, Istanbul, Turkey, pp.151-156.

Demirbas, G. (2000) *Port Privatisations,* Proceedings on Coastal Facilities and Ports, The 2nd National Maritime Congress, Undersecretariat of Shipping, Istanbul, Turkey, pp.149-157.

Japan International Cooperation Agency - JICA - and General Directorate of State Ports and Airports Construction in Turkey - SPAC (2000) *Report for the Study on the Nationwide Port Development Master Plan in the Republic of Turkey*, The Overseas Coastal Area Development Institute of Japan, Japan, pp.I-10.1-I-10.9.

Lloyd's List (1999) *Special Supplement: Turkish Shipping*, April, Lloyd's of London Press, London, pp.5-9.

The Working Group on Coastal Facilities and Ports (2000) *Report of the Working Group on Coastal Facilities and Ports*, Proceedings on Coastal Facilities and Ports, The 2nd National Maritime Congress, Undersecretariat of Shipping, Istanbul, Turkey, pp.105-130.

The World (1997) *Shipping*, Istanbul, Turkey, 30th June.

Turkish State Railways (2000) *Annual Statistics: 1995-1999*, Ankara, Turkey, p.72.

Yercan, F. (1999) Analysis of Privatisation Actions in the Maritime Industry in Turkey, in M. Ergun and J. Zurek (eds) *Strategic Approaches for Maritime Industries in Poland and Turkey*, Dokuz Eylul Publications, Izmir, Turkey, pp.127-135.

Yercan, F. and Roe, M. (1999a) *Port Privatisation in Turkey*, Proceedings, The 15th International Port Conference on Port and Transport Development in the Next Millennium, Alexandria, Egypt.

Yercan, F. and Roe, M. (1999b) *Shipping in Turkey*, Ashgate, Aldershot, UK, pp. 94-100.

Printed and bound by CPI Group (UK) Ltd, Croydon, CR0 4YY

21/10/2024

01777082-0003